ENVIRONMENTAL ACTIVITIES IN URANIUM MINING AND MILLING

A JOINT REPORT BY THE
OECD NUCLEAR ENERGY AGENCY

AND THE
INTERNATIONAL ATOMIC ENERGY AGENCY

NUCLEAR ENERGY AGENCY
ORGANISATION FOR ECONOMIC CO-OPERATION AND DEVELOPMENT

ORGANISATION FOR ECONOMIC CO-OPERATION AND DEVELOPMENT

Pursuant to Article 1 of the Convention signed in Paris on 14th December 1960, and which came into force on 30th September 1961, the Organisation for Economic Co-operation and Development (OECD) shall promote policies designed:

- to achieve the highest sustainable economic growth and employment and a rising standard of living in Member countries, while maintaining financial stability, and thus to contribute to the development of the world economy;
- to contribute to sound economic expansion in Member as well as non-member countries in the process of economic development; and
- to contribute to the expansion of world trade on a multilateral, non-discriminatory basis in accordance with international obligations.

The original Member countries of the OECD are Austria, Belgium, Canada, Denmark, France, Germany, Greece, Iceland, Ireland, Italy, Luxembourg, the Netherlands, Norway, Portugal, Spain, Sweden, Switzerland, Turkey, the United Kingdom and the United States. The following countries became Members subsequently through accession at the dates indicated hereafter: Japan (28th April 1964), Finland (28th January 1969), Australia (7th June 1971), New Zealand (29th May 1973), Mexico (18th May 1994), the Czech Republic (21st December 1995), Hungary (7th May 1996), Poland (22nd November 1996) and Korea (12th December 1996). The Commission of the European Communities takes part in the work of the OECD (Article 13 of the OECD Convention).

NUCLEAR ENERGY AGENCY

The OECD Nuclear Energy Agency (NEA) was established on 1st February 1958 under the name of the OEEC European Nuclear Energy Agency. It received its present designation on 20th April 1972, when Japan became its first non-European full Member. NEA membership today consists of all OECD Member countries, except New Zealand and Poland. The Commission of the European Communities takes part in the work of the Agency.

The primary objective of the NEA is to promote co-operation among the governments of its participating countries in furthering the development of nuclear power as a safe, environmentally acceptable and economic energy source.

This is achieved by:

- *encouraging harmonization of national regulatory policies and practices, with particular reference to the safety of nuclear installations, protection of man against ionising radiation and preservation of the environment, radioactive waste management, and nuclear third party liability and insurance;*
- *assessing the contribution of nuclear power to the overall energy supply by keeping under review the technical and economic aspects of nuclear power growth and forecasting demand and supply for the different phases of the nuclear fuel cycle;*
- *developing exchanges of scientific and technical information particularly through participation in common services;*
- *setting up international research and development programmes and joint undertakings.*

In these and related tasks, the NEA works in close collaboration with the International Atomic Energy Agency in Vienna, with which it has concluded a Co-operation Agreement, as well as with other international organisations in the nuclear field.

Publié en français sous le titre :

ASPECTS ENVIRONNEMENTAUX DE LA PRODUCTION D'URANIUM

PREFACE

Since the mid-1960s, with the co-operation of their members, the OECD Nuclear Energy Agency (NEA) and the International Atomic Energy Agency (IAEA) have jointly prepared the periodic report "Uranium Resources, Production and Demand". The report commonly known as the "Red Book" is published by the OECD. The seventeenth edition of the "Red Book" was published in 1998.

In 1996, the Joint NEA-IAEA Uranium Group which prepares the "Red Book", established a "Working Group on Environmental Issues in Uranium Mining and Milling". This was done in response to an initiative from the Swedish Minister of Environment Ms. Anna Lindh, supported by both the IAEA and NEA, encouraging the Joint NEA-IAEA Uranium Group to broaden its mandate to foster the exchange of information on environmental effects and environmental technologies associated with uranium mining and ore processing.

To obtain an overview of this theme, the Working Group sent out a questionnaire requesting information about current activities and interests of participating Member countries/states with uranium-related activities. It also reviewed the relevant IAEA and NEA programmes. The results of this survey form the basis of the report on "Environmental Activities in Uranium Mining and Milling" presented to the Uranium Group in the fall of 1997. Following review, the Joint NEA-IAEA Uranium Group concluded that the report was of sufficient interest to be published as a complement to the seventeenth edition of the Red Book.

The report is not intended to be a comprehensive review of all related environmental issues for all countries. As environmental and safety aspects related to uranium mining and milling is a broad subject, the questionnaire asked for submissions from each country based on what participants considered to be of importance. The report therefore includes information related to those activities judged to be important by the contributing countries.

The options expressed in this report do not necessarily reflect the position of Member countries or international organisations. This report is published on the responsibility of the Secretary-General of the OECD.

Acknowledgement

The Working Group and the Joint NEA-IAEA Uranium Group would like to acknowledge the co-operation of all the organisations (see Annex 4) which submitted information for this report.

TABLE OF CONTENTS

ANNEXES

EXECUTIVE SUMMARY

This report on "Environmental Activities in Uranium Mining and Milling" presents an overview of environmental activities related to uranium production. The profile of activities and concerns are based on survey responses from 29 countries and a review of relevant activities of the International Atomic Energy Agency and the OECD Nuclear Energy Agency. It also provides an overview of the reported interests of specialists working in the field, including environmental impact assessment, emissions to air and water, work environment, radiation safety, waste handling and disposal, mine and mill decommissioning and site restoration, and the regulation of these activities.

The report reflects the increasing awareness in all countries of the need for environmental protection. For several years large programmes have been underway in several countries to clean up wastes from closed mines and mills. Many of these sites, particularly the older ones, were brought into production, operated and closed when little was known about environmental effects. At the time, little concern was given to the resulting environmental impacts. Currently, planning for and conducting uranium mine closure and mill decommissioning, together with site clean-up and restoration, are of almost universal concern. Mine closure and mill decommissioning activities have been or are being conducted in most of the countries with a history of uranium production. Information about several mine closures and mill decommissioning projects is included in this report.

For new projects, managers must prepare and submit an environmental impact assessment to regulatory authorities. This normally consists of detailed planning for the life of the project, including relevant safety and environmental aspects. These plans and assessments are scrutinised by the responsible authorities before decisions are taken to allow – or not allow – the new project to proceed. The environmental assessment process may involve public hearings that provide for discussion of socio-economic impacts and the concerns of stakeholders in the communities affected by the project. In particular, information is given in this report on new projects being developed in Australia and Canada that must successfully complete environmental assessment programmes.

Measures taken to limit emissions to air and water, and the new rules for radiation protection of both project personnel and the general public are discussed. These measures, implemented as the "Basic Safety Standards", are becoming increasingly stringent to provide appropriate radiation protection. The new "Joint Convention on the Safety of Spent Fuel Management and on the Safety of Radioactive Waste Management" is expected to be applied to wastes from uranium mining and milling in the future. New technology such as "non-entry mining" has been developed to make possible the mining of deep, high-grade ores in Canada. In this system mine personnel carry out routine ore extraction without entering the ore-zone. In addition, the technology for closing mines and decommissioning mills is also advancing, and improved methods are being developed for long-term isolation of contaminated wastes from uranium mines and mills.

Nuclear power makes a major contribution to sustainable development as it does not generate greenhouse gas emissions and has a low environmental impact. For several decades the uranium mining and milling industry has been improving the efficiency of its operations while reducing impacts on both humans and the natural environment in general. This includes the reduction of short-term impacts as well as long-term effects of waste disposal. There are sufficient known uranium resources to sustain nuclear power for many years without compromising the needs of future generations. This will only be possible, however, as long as the managers and operators of uranium production projects keep implementing responsible planning, operation and closure, to minimise environmental and health impacts.

I. INTRODUCTION

There are increasing environmental concerns related to the mining and milling of raw materials used in the production of energy. This interest reflects a renewed effort to use integrated assessments in the evaluation of technological processes that could potentially contribute to sustainable development.

The only large-scale civilian use of uranium is for nuclear fuel. There are sufficient known uranium resources to sustain nuclear power for many years without compromising the needs of future generations. The sustainability of this industry will only be possible, however, as long as the managers and owners of uranium production facilities implement responsible planning, operational and closure programmes, to minimise environmental and health impacts.

In the case of uranium mining and milling, environmental issues have become increasingly important in the last few decades due to important developments affecting the uranium industry. These developments include: the large number of uranium production facilities which have been taken out of operation recently; the stricter requirements being imposed on new facilities by many countries in the form of environmental clearance approvals; and, the restoration and reclamation measures that are being considered for many old sites which were abandoned at a time when provisions for decommissioning and rehabilitation were not sufficient.

This report provides an overview of environmental activities related to uranium production. The profile of activities and concerns is based on survey responses from 29 countries and a review of relevant activities at the International Atomic Energy Agency and the OECD Nuclear Energy Agency. The report also discusses environmental and safety aspects related to closure and remediation of formerly utilised sites; the operation, monitoring and control of producing sites; and the planning, licensing and authorisation of new facilities.

The report consists of 5 major sections: an introduction, an overview of the environmental and radiation protection aspects related to uranium mining and milling, a list of related activities at the IAEA and the NEA, a summary of the activities reported by countries, and the national submissions. The section on environmental protection and radiation safety (Section II) includes: overviews of radiological and non-radiological impacts; measures currently being implemented to ensure the adequate protection of workers, the general public and the environment; a list of relevant international standards; and, a review of technical aspects and typical emissions from uranium mines and mills. The section ends with a brief discussion of world trends in uranium related environmental activities and the corresponding implications to sustainable development. A list of the most relevant activities by the IAEA and the NEA is provided in Section III. In addition to co-operative projects, the section includes the most important recent publications by these two international organisations. Annex 2 of this report complements this section by providing a more complete list of activities and publications by the IAEA. Section IV provides a summary of the major activities reported by countries. Activities are summarised in the areas of: environmental impact assessments, emission control, radiation safety, decommissioning and restoration, regulations and others.

The national reports (Section V) include information on environmental activities provided by 12 OECD countries and 17 non-OECD countries. The reported information varies from country to country in scope and level of detail and no attempt has been made to standardise the submissions as to follow a particular format or style.

II. ENVIRONMENTAL ASPECTS RELATED TO
URANIUM MINING AND MILLING[1]

The environment is composed of interacting systems including biological elements, fauna and flora, physical elements, atmosphere, land and water, and social and cultural elements. Impacts can be defined as those effects that alter the existing system either temporarily or permanently. The definition and measurement of impacts are highly dependent on location, country, and social and economic factors.

Figure 1 illustrates the flow of materials and wastes in a generic facility. Each facility produces a product from raw materials by using resources (energy, water, land) and reagents. The facility also produces wastes in the form of environmental releases or effluents to air, water, and the soil. Some of these wastes (i.e. mill tailings) are normally collected and contained in engineered structures or controlled areas. These radiological and non-radiological releases may have impacts on humans (workers and the general public) and other living organisms, and on the physical environment such as changes in water, sediment or air quality. Because of the long-lived nature of some of the radionuclides released in the nuclear fuel cycle, concentration and bio-accumulation in all food chains must be taken into account where appropriate when evaluating environmental impacts.

Mining and milling of ores usually involves the disruption of the land surface, and may impact both surface and underground water bodies. Therefore the environmental impacts are potentially broader and more diverse than for other steps of the nuclear fuel cycle (i.e. conversion, enrichment, fuel fabrication, fuel reprocessing, etc.). Annex 1 provides an evaluation of the operational impacts of both uranium mines and mills including an illustrative example. For additional information on the facilities, including release data, the reader is referred to the inter-agency DECADES (Databases and Methodologies for Comparative Assessment of Different Energy Sources for electricity generation) project database at the IAEA. Several reports are also available which describe the effect of the overall nuclear fuel cycle on the environment [1,2,3,4,5].

Within the nuclear fuel cycle, nuclear safety analysis can be used in environmental assessments specifically addressing accident occurrences. This report focuses on the normal operation of uranium production facilities and only briefly refers to the consideration of accidents.

Overview of radiological impacts

During normal operation of uranium production facilities small amounts of radioactive materials released into the environment may result in some radiological impact on workers, the general public and the environment.

---

1. A large part of the information contained in this section was derived from the 1996 IAEA report on *Health and Environmental Aspects of Nuclear Fuel Cycle Facilities*, IAEA-TECDOC-918.

Radiological impacts on humans in the nuclear fuel cycle are stated in terms of dose. Doses to the public may be calculated using a method known as Pathways Analysis. Figure 2 shows a typical set of pathways used in doing this calculation. In order to perform dose calculations, a knowledge of the source term of the transfer pathways in the human food chain and the transfer parameters between each compartment in that food chain is required. Occupational doses to workers on the other hand are usually assessed using direct measurements rather than modelling. International organisations such as the United Nations Scientific Committee on the Effects of Atomic Radiation (UNSCEAR) have done much to systematically collect, analyse, and present information in this area. Information on radiological impact analysis may be obtained from the 1993 report of the UNSCEAR [6]. In its 1993 report to the United Nations General Assembly, the UNSCEAR evaluated the normalised collective doses to the public and annual occupational exposures to monitored workers from the various steps of the nuclear fuel cycle. The collective dose concept allows consideration of the impact on a local or regional level and on a global level. Data showing the collective dose resulting from mining and milling of uranium are presented in Tables 1 and 2.

Figure 1. **Material flow and impacts**

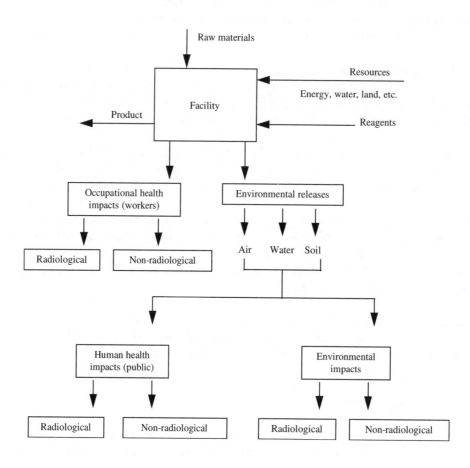

The normalised collective effective dose shown is based on a model with a population density of three people per km^2 in the vicinity of mills. However, much of the world's uranium is produced in areas with a lower population density than this. The doses are, therefore, likely to be overestimates.

Figure 2. **Typical set of pathways for analysis [1]**

Schematic representation of atmospheric pathways

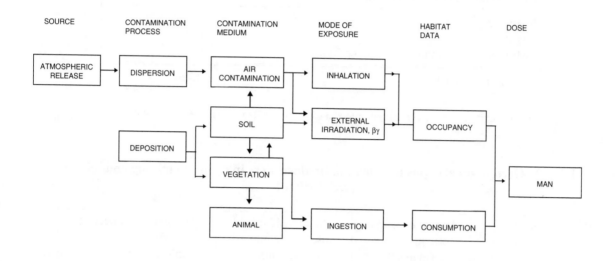

Schematic representation of aquatic pathways

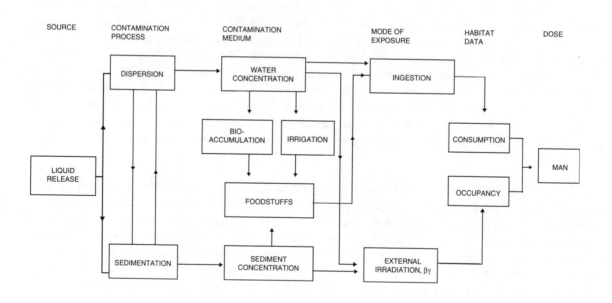

Table 1. **Normalised collective effective dose to the public from uranium mining and milling [6]**

Source: normalised collective effective dose to the world public (manSv per 200 tU/year*)	
Local and regional component	
Mining	1.1
Milling	0.05
Mine & mill tailing (releases over 5 years)	0.3
Total (rounded)	**1.45**

* Quantity of uranium required to feed a Light Water Reactor (LWR) of 1 GWe.

Table 2. **Occupational exposures to monitored workers in uranium mining and milling, 1985-1989 [6]**

Occupational category	Annual collective effective dose (manSv)	Annual effective dose per monitored worker (mSv)	Normalised collective effective dose (manSv per 200 tU/year)
Mining	1 200	4.4	4.3
Milling	120	6.3	0.44

For the entire fuel cycle, the local and regional normalised collective effective doses, which are effectively received within one or two years of discharge, amount to 3 manSv per GWe/year and are principally due to routine atmospheric releases during reactor and mining operations. The global collective dose for the nuclear fuel cycle has been calculated to be about 200 manSv per GWe/year. This collective dose represents the total radiological impact on the population of the earth resulting from the nuclear fuel cycle. As a reference, the typical annual effective dose of 2.4 mSv from natural sources results in an annual collective dose to the world population of 5.3 billion people of about 13 million manSv. Figure 3 presents the relative radiological impact from all sources expressed in percentages of global average annual doses. The estimates are based on the 1993 UNSCEAR study [6].

In the UNSCEAR study, the collective dose from radon emitted from uranium mill tailing areas was estimated to be 150 manSv per GWe/year by using estimated values in a calculation from a theoretical site. Recently, a study with real data on radon release rates, tailings surface areas, population densities and air dispersion factors including reduction in connection with distance, comprising real mining sites which produce 80% of the present world uranium production was presented at the IAEA [7]. This new study shows a much lower collective dose of 1 manSv per GWe/year. UNSCEAR is now considering this new study in the preparation of its next report.

Concerning the protection of fauna and flora, recent analyses based on generalised information and conservative assumptions, have shown that radiation levels implied by current radiation protection limits for the public, are generally adequate to protect other species (plants and animals) and that only the combination of specific ecological conditions such as the presence of rare or endangered species and specific stresses may require site specific analyses [8]. However, additional work is underway on this issue.

The environmental protection strategy for nuclear fuel cycle facilities is to operate without exceeding the national and international radiation limits for humans. As part of this strategy environmental monitoring and studies are carried out during all phases of the facility's operation.

Figure 3. **Estimated shares of global average annual doses based on UNSCEAR 1993 [6]**

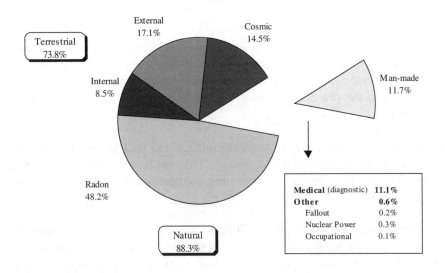

Overview of non-radiological impacts

Many different chemicals and types of equipment are used in mining and milling facilities. In most cases they are the same as those used in other industries. Thus the effects of using these chemicals and equipment are also much the same. Mitigative measures consist of proper codes of practice, appropriate waste treatment systems and a good safety culture. Therefore, the environmental protection strategy for non-radiological substances released to the environment by nuclear fuel cycle facilities are generally the same as for non-nuclear industries. Environmental monitoring and studies carried out for radiological contaminants in nuclear fuel cycle facilities would normally include monitoring and studies of non-radiological contaminants. Table 3 shows potential radiological and non radiological sources of environmental impacts and mitigation from uranium mining and milling facilities.

Table 3. **Potential radiological and non radiological sources of environmental impacts and mitigation from mining and milling**

Source	Mitigation
Temporary land disruption	Reclamation/proper practices
SO_x emissions	Scrubbers
Silica dust	Ventilation/filters
Radon and radon progeny	Ventilation, water or soil covers
Heavy metals	Effluent treatment
NH_3	Removal
Acid production	Neutralisation/impermeable covers
Tailings	Tailings management facility designed to retard migration of radionuclides and reduce radon emissions

Radiation safety

Most countries have legislation to control mining activities, as well as general legislation for occupational health and safety. Uranium mining and milling operators take precautions, including the provision of health and safety training, to limit any impact on the health of workers and to avoid accidents. Appropriate protective equipment is provided for those working in the facilities, and safety audits are conducted to ensure that equipment is being properly used and that work is being done in a safe manner. To increase public safety, access to areas where uranium mining and milling operations are taking place is restricted.

The radioactivity of uranium and associated radioactive elements implies the use of specific safety precautions in addition to those implemented in similar workplaces where there is no radiation risk. The extent of precautions required to control the radiation dose of mine and mill workers is, for the most part, a function of the ore being mined. The intensity of radiation increases as the uranium content in the ore increases.

Radiation levels around many types of uranium mining and milling facilities are relatively low, and in most cases, only a few times natural background. They decrease rapidly as the distance from the operations increases. The risk to workers is low as long as appropriate measures are taken to maintain exposures within allowable limits.

Monitoring of both the radiation levels at facilities and the exposure of workers is an essential part of radiation protection procedures. Where workers are likely to receive radiation doses that are a significant fraction of the regulatory limit, individual records of exposures are kept. Personal dosimeters may be used to monitor the exposure of each worker. In addition, a programme of medical surveillance is implemented.

In mining operations where the ore grade is high, the mining technique and operating procedures are selected to reduce and control the radiation exposure of workers. This is particularly true for underground mines. In some cases, non-entry mining methods may be required. In this type of operation remote controlled equipment is used to extract ore which is operated by personnel located outside of the ore zone.

When high grade uranium is milled, radiation shielding may be required to control the exposure of equipment operators. Once the milling process separates the uranium from the rest of the ore, which includes nearly all the radioactive decay products, very little radiation is emitted by the uranium concentrate. Consequently, the likelihood of workers receiving significant direct radiation exposure from uranium concentrate is greatly reduced.

Radon decay products are, if inhaled, a potential source of radiation exposure. The most effective way of limiting radiation exposure from this source is by providing adequate ventilation. Such ventilation is required wherever personnel work in a closed area with uranium ore. This is a particularly important consideration in underground mines.

Airborne materials such as radon are diluted in the open air and rapidly disperse to background levels. Therefore, radon from uranium mining and milling operations presents no health hazard for the general public as long as they are restricted from entering or living in the immediate vicinity of the facilities.

Uranium dust in the air may be inhaled by workers. Where dust is a potential problem, wetting of the ore is the most commonly used control mechanism. In areas where high concentrations of dust may be present, such as facilities where uranium concentrate is packaged, personnel are required to wear respirators.

Supplying clothing and a programme of personal hygiene, such as providing washing and shower facilities, is usually sufficient to mitigate the risk of contamination of personnel.

Environmental protection

Increasingly many countries require that environmental impact assessments (EIAs) be conducted prior to development of proposed uranium mining and milling operations. In preparing such assessments, mine and mill operators identify the actions they intend to implement to limit impacts. Acceptance of the proposed actions are subject to the approval of governmental regulatory agencies. Regulatory agencies monitor the activities when the facilities are in operation. An introduction to the EIA process for uranium production, together with some national examples, is given in a recent IAEA publication [9].

The two principal pathways by which contamination may reach the environment from uranium mining and milling operations are air and water. Frequent sampling is carried out in the vicinity of uranium production facilities, and monitoring of the affected environment is undertaken, to ensure that environmental impacts are adequately controlled within allowable limits. Such monitoring normally includes sampling of air, soil, water, plants and animals, including fish.

Research has been carried out by the uranium industry to develop operating strategies that meet regulatory requirements, limit long-term impacts on the environment, demonstrate that facilities can be successfully decommissioned, and show that waste sites can be safely closed.

Dust created by mining and milling activities, or blown by the wind from ore stockpiles, is a potential source of environmental contamination. Methods used to prevent spreading of wind blown dust may include use of enclosures, and the application of cover materials or water, to ore stockpiles, tailings and other loose material. Where ventilation carries significant amounts of uranium bearing dust, scrubbers and filters are used to remove the dust before the air is discharged to the environment.

To reduce water use, the quantities of fresh water used in milling operations are limited so far as is practical. While some operations require clean water, where possible, extensive use is made of recirculated water, especially at sites where the water supply is limited. Water contaminated during mining and milling operations, that cannot be reused, is handled as waste. This water may not be discharged directly into the environment. In some areas, evaporation from holding ponds may be sufficient to eliminate the need to discharge water. In some situations, engineered barriers, such as berms and dams, are installed to prevent contamination of surface or groundwater by uranium mining and milling operations. These barriers serve to isolate the uranium mine or mill from the surrounding environment.

Waste management

There are three main types of waste arising in mining and milling operations: mine wastes, milling wastes and waste water. Each type of waste is subject to an appropriate management strategy.

Mine waste consists primarily of waste rock and low grade ore that must be removed to access the ore. For the most part this material presents essentially no risk of environmental contamination. It may then be disposed of on the surface adjacent to the mine. In some cases, waste rock contains minerals, including sulphides, that may be leached by water passing through waste piles. Oxidation of sulphide minerals is the process which contributes most to acid drainage and to the mobilisation of metals from mine waste to the environment. The waste rock may also include concentrations of uranium that are elevated above the normal background, but do not justify processing and recovery. When this is considered likely, steps should be taken to ensure that neither the rock nor leachate leaves the site thereby preventing contamination of either the surface or ground water.

The primary milling wastes consist of tailings. The overriding long-term environmental issue associated with uranium mining and milling is the effective isolation from the accessible environment of long-lived radionuclides that readily migrate from uranium mill tailings into life-support systems and food chains and have significant biological half-lives or residence times. To prevent undesirable constituents in the tailings from leaching into the environment an engineered containment should be constructed for the long-term disposal of the material. This provides for tailings management during and after the life of the facility, thereby limiting the long-term environmental impact. In the past some operators disposed of tailings at sites where little or no consideration was given for containment of the material. While this practice is no longer acceptable, some large environmental projects involve evaluation, stabilisation and close-out of these formerly used sites.

At some uranium mining and milling operations tailings are disposed of by placing them in an open pit following the completion of mining. An improvement of this disposal system was developed in the province of Saskatchewan, Canada, where the "pervious surround" tailings disposal system is used. Use of this technique involves placing a porous sand layer between the pit wall and the tailings. This helps limit releases to the environment as the water flow through the tailings is reduced to very low levels. Today the preferred option, where appropriate, is to return the tailings to the void where the ore was excavated.

Mill tailings are generally disposed of as a slurry, i.e. a water and solid mixture. Following placement in a tailings management facility, the solids tend to settle and the water can be decanted. This water contains a high concentration of radium, and may contain other potential contaminants. It must therefore be treated before it can be released to the environment.

Waste water from mining and milling operations that is not recycled to the mill must be either contained at the mill site or treated before discharge to the environment. Standards have been established for maximum concentrations of specific contaminants in discharged water. Generally, radium, a decay product of uranium, is one of the critical radiological contaminants for determining whether treated water meets the limits for release to the environment. Radium is relatively easily removed from water when appropriate technology is used. There are several other critical non-radiological trace elements that may present more of a problem than radium.

In situ leaching

As no ore is brought to the surface during in situ leach (ISL) mining, concerns relating to radiation protection and waste management are greatly reduced. Surface disturbances are minimised, there are no tailings, and therefore environmental impacts are much less than for conventional mining. The most important operational considerations are routine in-plant radiation health concerns, as well as monitoring and controlling ground water conditions in and around the mined out ore body.

International regulations, standards and guidelines

The regulation of uranium production activities is primarily controlled by environmental legislation in individual countries. However, there are a growing number of international conventions that impose environmental obligations on signatory governments. The IAEA facilitates the establishment of international conventions that address environmental issues which may relate to uranium production facilities such as:

- The Convention on Environmental Impact Assessment in a Transboundary Context.

- The Convention on Access to Information, Public Participation in Decision Making and Access to Justice in Environmental Matters.

- The Convention on the Protection of the Environment through Criminal Law.

The Convention on Environmental Impact Assessment in a Transboundary Context applies to both nuclear installations and radioactive waste management facilities. The Convention provides for certain rights and duties of contracting parties when the environmental impact of an activity has a transboundary effect and for certain procedures to be followed when considering the environmental impact of a given project. The Convention was signed in 1991 by 55 countries and ratified by 21 countries, all of which are IAEA Member states. The Convention entered into force in October 1997.

The Convention on Access to Information, Public Participation in Decision making and Access to Justice in Environmental Matters applies to both nuclear installations and radioactive waste management facilities. Its objectives are to contribute to the "protection of the right of every person of present and future generations to live in an environment adequate to his or her health and well-being" and to guarantee "the rights of access to information, public participation in decision-making, and access to justice in environmental matters". The Convention was signed on 25 June 1998 by 35 States, all of which are IAEA Member states. Adoption of the Convention was combined with the Fourth Ministerial Conference, "Environment for Europe" Declaration of 52 ECE Ministers for the Environment. The Convention has not yet entered into force.

The Convention on the Protection of the Environment through Criminal Law makes the unlawful discharge, emission or introduction of a quantity of ionising radiation into air, soil or water an intentional offence punishable and treated as such by national law if signatory to the Convention. The Convention was signed on 19 May 1999 by 9 States, all of which are IAEA Member states.

Other international conventions that may effect uranium mining and milling include: the 1972 World Heritage Convention, the 1989 Biodiversity Convention and the 1997 Kyoto Protocol. A recent report by the United Nations Environmental Programme (UNEP) describes the changing nature of environmental regulation in mining facilities [10].

An overview of international standards and guidelines applicable to uranium mining and milling is given as follows:

Industrial safety in the working environment

International standards for the conventional work environment applicable to uranium milling operations are published by the "International Labour Organisation" (ILO). These include ILO Convention No. 170 "Concerning safety in use of chemicals at work", with the corresponding

Recommendation No. 177, and the ILO Convention No. 174 "Concerning the prevention of major industrial accidents", with the corresponding recommendation No. 181.

Basic safety standard for protection against ionising radiation

The two most important international standards for radiation safety in uranium mining and milling are published by the IAEA:

- The "International Basic Safety Standards for Protection against Ionising Radiation and for the Safety of Radiation Sources", Safety Series No. 115, 1996 Edition.

- The new "Joint Convention on the Safety of Spent Fuel Management and on the Safety of Radioactive Waste Management".

The "International Basic Safety Standards for Protection against Ionising Radiation" was jointly sponsored and agreed to by a number of international organisations: FAO, IAEA, ILO, OECD/NEA, PAHO and WHO, and published in 1996. These standards were established based on recommendations from the International Commission for Radiation Protection, (ICRP). The new Basic Safety Standard (BSS) replaces an earlier version: Safety Series No. 9: *Basic Safety Standards for Radiation Protection*, 1982 Edition, which is obsolete.

In the new BSS, the effective dose limit for workers in uranium mining and milling is 20 mSv per year averaged over 5 consecutive years, including an effective dose of 50 mSv in any single year. This is more restrictive than the limit published earlier which was 50 mSv/year without any limitation for a five-year period. In the new BSS, the dose limit for the general public is 1 mSv per year, as compared with 5 mSv per year, in the previous version.

The new BSS also contains prescriptions to limit radiation doses from natural radiation sources. Most uranium-producing countries have made a decision to implement the new BSS. The implementation will take place within a few years. For example, the European Union adopted its own version of the BSS in 1996, and will implement it in all member countries by the year 2000.

Joint convention on the safety of spent fuel management and on the safety of radioactive waste management

The new "Joint Convention on the Safety of Spent Fuel Management and on the Safety of Radioactive Waste Management" is applied to uranium mining and milling wastes, and to corresponding decommissioning and closure activities. This Joint Convention was approved by an International Diplomatic Conference in Vienna, 5 September 1997. The Convention has been open for signing and ratification since 29 September 1997. The Convention will enter into force when 25 countries, 15 of which with at least one nuclear power plant, have ratified the text.

The Joint Convention contains a chapter on "Safety of Radioactive Waste Management" including treatment of operational and institutional measures after closure. There is also a chapter on "General Safety Provisions", including legislative and regulatory framework, regulatory body, quality assurance, operational radiation protection, emergency preparedness, responsibility of the license holder, human and financial resources, and decommissioning. Another article addresses transboundary movements of waste.

IAEA safety standards and safety guides

The IAEA has published a number of safety standards and safety guides including:

- Safety Series No. 26: Radiation Protection of Workers in the Mining and Milling of Radioactive Ores, 1983 Edition. Code of Practice and Technical Addendum sponsored by IAEA, ILO, WHO.

- Safety Series No. 85: Safe Management of Wastes from the Mining and Milling of Uranium and Thorium Ores, 1987 Edition. Code of Practice and Guide to the Code.

- Safety Series No. 111-F: The Principles of Radioactive Waste Management: A Safety Fundamental, 1995 Edition.

- Safety Series No. 111-S-1: Establishing a National System for Radioactive Waste Management: A Safety Standard, 1995 Edition.

The first two safety series documents supersede earlier codes of practice published in 1976. Other documents include:

- Safety Series No. 82: Application of the Dose Limitation System to the Mining and Milling of Radioactive Ores, 1987 Edition.

- Safety Series No. 90: The Application of the Principles for Limiting Releases of Radioactive Effluents in the Case of the Mining and Milling of Radioactive Ores, 1989 Edition.

- Safety Series No. 95: Radiation Monitoring in the Mining and Milling of Radioactive Ores, 1989 Edition.

European Union regulations

In the European Commission, Council Directive 89/391/EEC introduces measures to encourage improvements in health and safety of workers which are also applicable to uranium mining and milling workplaces. The Council Directive 96/29 EURATOM introduces similar regulation for radiation safety. In addition, Council Directives 74/326/EEC, 92/58/EEC, 92/104/EEC and 89/391/EEC provide regulations to protect the health and safety of mine and mill workers. The Social Europe 3/93 contains the principal texts. Council Directive 96/29 EURATOM includes Basic Safety Standards to protect the health of workers and the general public against risks from radiation. The implementation in Member countries will be in the year 2000.

Mining

Technical aspects

Uranium is widely distributed in the earth's crust and oceans where it has an average abundance of two parts per million (ppm) and five parts per billion (ppb), respectively. The average concentration required for economic recovery depends, among other things, on the market price of uranium. In recent years the uranium market price has been severely depressed and only higher grade deposits capable of lower cost production have continued to operate. Historically, uranium has been economically recovered using conventional production where ores contain average grades of about

0.1% uranium or more. In most cases in the western world where average grades are between 0.01% and 0.1% uranium [i.e. 100 to 1 000 parts per million (ppm)], uranium is recovered as a by-product of other mineral commodities. In Namibia the conventional Rossing mine and mill produces uranium from ores with an average grade of about 0.04% U. In Canada, deposits with up to 20% uranium are being developed for production.

The ores containing uranium bearing minerals are usually mined by conventional open pit or underground methods depending on the geological condition of the ore, such as deposit size, ore grade, depth and ground condition [11]. In general, the open pit method is employed when the ore body lies close to the surface under an overburden which can be easily and economically removed. Underground mining is typically used for ore bodies at depths greater than 100 m. This operation produces less waste rock than open pit mining. Conventional mining produced 79% of the world's uranium in 1996, 39% by open pit and 40% from underground mines [12].

Some non-conventional methods, such as in situ leaching (ISL) and heap leaching, are also used for uranium production. However, heap leaching is of minor importance. ISL mining requires a porous ore body (sandstone) saturated with groundwater and confined between relatively impervious layers. The ore is left in the ground and a leaching solution (either alkaline or acidic, plus an oxidant) is injected into the ore body through wells. The solution percolates through the ore where it oxidises and dissolves the uranium. The uranium bearing solution is then recovered by pumping. Uranium is extracted from the solution in a surface facility using ion exchange technology similar to those methods employed in some conventional ore processing plants. ISL mining is being used to extract uranium at a maximum depth of about 300 m in the USA, and about 550 m in Kazakhstan. ISL mining produced 13% of 1996 world uranium production [12].

In most cases alkaline ISL systems, using bicarbonate solutions and dissolved oxygen as an oxidant, are considered to be more environmentally acceptable than are conventional mining operations. However, ISL mining can result in significant groundwater contamination, when a sulphuric acid system is used in an inappropriate location. Groundwater restoration following acid ISL mining is thought, in most cases, to be more difficult to achieve.

Heap leaching is a similar process whereby broken ore is leached, either underground or on an impervious membrane located on the surface. Leaching is by percolating sulphuric acid (or less commonly, alkaline solutions), through the heap and recovering the uranium solution. Heap leaching is usually applied to low-grade ore produced from a conventional mining operation, which could not be economically milled using conventional processing. Currently, the amount of production from heap leaching is small.

In addition, uranium is recovered as a by-product of other mineral processing, such as phosphate fertiliser, as well as mining of gold and copper, and some other metals. This represented about 7% of 1996 world production (not including South African production which is normally included with conventional underground production) [12].

The quantity of ore required to produce nuclear fuel depends on the average grade of the ore. In recent years the grade has usually ranged between 0.1% and 2.5% U. Production of about 200 tonnes uranium annually corresponds to a mining rate of between 8 000 to 200 000 tonnes ore. The higher grade deposits now being developed will require a much lower rate of ore extraction.

The amount of temporarily committed land use for open pit mining is currently estimated to be 25 hectares per 200 annual tU. However, this figure is expected to decrease as lower grade mines are exhausted and higher grade mines are developed. About one hectare of land is permanently committed to institutional control [4]. During open-pit mining operations, earth overburden above the orebody and barren rock are produced. The quantity of the waste rock removed is estimated to be around 10^6 tonnes per 200 tU of annual production [2].

Underground mining requires the disturbance of relatively small areas of land, primarily for waste rock piles. During ISL mining, land surface areas are only temporarily used and the disturbance from the operation is small.

Open pit mining produces a greater surface disturbance than either underground or ISL mining. Surface land disturbances must normally be remediated following open pit mining. Consequently, the amount of land permanently committed is likely to be less than 1 ha per 200 tU of annual production.

Releases

The releases from uranium mining are, for the most part, similar to releases from conventional mines. They are, in addition to typical releases, radon and radon progeny, and radioactively contaminated water, dust, and others.

Radon (primarily radon 222) and radioactive dust are released to the atmosphere when an ore body is exposed and broken during mining operations. Short-lived radon progeny, resulting from the decay of radon, are a major source of radiation exposure for uranium mine workers, particularly in underground mines. Ventilation is used in underground mines to remove radon and thereby limit the exposure to its progeny. However the expelling of the radon and its progeny from underground mines results in dispersal of these radionuclides into the environment. At ISL mines radon gas is dissolved in the uranium bearing solution that is pumped from the ore body. This radon may be released if the solution is exposed to the environment in tanks or ponds. The average radon release in mining is 75 TBq per 200 tU/year [6].

Water contaminated with radioactive or other materials is produced by the dewatering of underground and open pit mines, surface water runoff from, and seepage through, the waste rock piles and ore stockpiles. Contaminated water is produced in association with both routine ISL operations and during ground water restoration activities. The radioactivity of this water is generally derived from the dissolution of soluble uranium, thorium, radium and lead ions. The water may also be contaminated with various heavy metals such as arsenic, selenium, and nickel. In cases where pyrite is present in the ore, the generation of acid requires neutralisation of the water as part of the treatment process, if it is discharged to the environment. Acid generation is a concern of all types of mining as the acid dissolves and increases the mobility of heavy metals. It also mobilises the radionuclides present in uranium ores. Blasting may also add nitrites, nitrates, and ammonia to the mine water.

In many cases, contaminated water is collected and may be recycled for use in the mill process. This presents a means of reducing the amount of contaminated water, while recovering the small amount of uranium that may be present in this water. A consequence from the mine water recycling is that only one treatment facility and one point of discharge are required, thus simplifying the monitoring of releases. In the treatment facility the effluent is neutralised where necessary, and chemicals and flocculants are added. The treatment facility comprises one or more evaporation ponds through which the treated effluent is passed to provide for the precipitation of contaminants before release to the environment.

Dust originating at exposed ore stockpiles may be controlled by applying water or installing wind resistant covers. It may also be necessary to apply water to roads to reduce dust.

As in any industrial undertaking, there will be possible releases of other substances. Some examples are fuel oils, contaminated solid waste and regular landfill material. Where these materials are contaminated with radionuclides, they are disposed of in the tailings area. However, this practice is not allowed in some countries, and it is not recommended by the IAEA. Other appropriate landfill arrangements are made for wastes with no radioactive contaminants. Local environmental regulations and requirements normally provide for disposing of these materials.

After termination of operations, most jurisdictions require some level of restoration during mine closure and mill decommissioning. Decommissioning requirements may include restoration of aquifers involved in ISL mining where there is a risk of contaminating aquifers used for drinking or other beneficial purposes.

At the end of mining it is usual that some mineralization may be exposed in the walls and floor of an open pit. This may result in the ongoing release of contaminants such as radon, other radionuclides, and heavy metals. However, it is common for mined-out pits to be flooded or some other form of remedial action to be taken to limit these releases. In some cases, mined-out pits are used for mill tailings disposal. In most cases, the amount of radon release from underground mines is much less after mining ends. Forced ventilation is terminated, the mine is flooded and mine openings are sealed.

The potential exists for contamination of adjacent aquifers during ISL mining. The requirements for having natural physical barriers (i.e. geological confining layers), plus mandated monitoring, makes adjacent aquifer contamination unlikely. Any contamination that might occur can be relatively easily remediated provided appropriate, timely actions are taken.

If open pit or underground mines extend below the water table, groundwater must be removed to permit mining operations. This can result in water extraction and temporarily lowering of the water table. The effect is generally limited to the immediate vicinity of the mine operations. The water table usually returns to its normal level after pumping is discontinued.

Upon completion of mining and milling operations, the facility must be decommissioned [13]. Decommissioning involves the administrative and technical actions taken to allow removal of some or all of the regulatory controls. Adequate steps must be taken for the health and safety of workers and members of the public and protection of the environment. Decommissioning plans are subject to approval by government regulatory agencies. Decommissioning and restoration are generally based on the principle that the area should, where possible, be returned to the condition and state in which it was before the operations commenced. When facilities are closed, monitoring is required for an extended period to ensure that there is unlikely to be a significant long-term environmental impact.

As part of closure, wastepiles which could be a source of heavy metal and radionuclide contamination, are recontoured and re-vegetated to limit water infiltration, radon release and wind erosion. In addition, any precipitates from the treatment of water must be disposed in an appropriate manner, as they may be a possible source of future contamination. In some cases, part of the underlying material may be disposed of in a tailings impoundment provided one is located at the site.

Accident considerations

The most likely accidents related to uranium mining operations that may have environmental consequences is the unplanned release of contaminated water from a pipeline or rupture of slurry transfer systems. The sudden failure of a tailings impoundment may also cause negative environmental impacts. In general, monitoring systems are in place to ensure that such releases are detected quickly, in order that the consequences may be limited and remedial action promptly undertaken.

Milling

Technical aspects

Uranium ore is usually processed close to the mine to limit transportation costs. The mine to mill distance is usually within 5 to 10 kilometres. However, in some cases, uranium ore (and/or slurry concentrate) may be transported several hundred kilometres by truck or rail to the mill for processing. In all cases, operational procedures are implemented to prevent release of these materials to the environment during their transfer and transport. Procedures for responding to accidental spills are also established. However, the location of the mill is primarily dependent on the availability of an environmentally acceptable tailing disposal site. The typical process for the extraction of uranium consists of crushing and grinding of the ore, followed by chemical leaching. Sulphuric acid leaching is the more common method. However, some mills use alkaline leaching when the ore contains limestone or similar basic constituents which would consume excessive quantities of acid. The uranium solution is purified and concentrated by ion exchange, and/or solvent extraction technology. The uranium is then precipitated from solution, filtered and dried to produce a concentrate, known as yellowcake. Yellowcake contains between 60% and 90% uranium by weight.

It is estimated that currently a mill facility requires an average of about 4 ha of land area per 200 tU/year [4]. About 75% of this land is devoted to an impoundment for the permanent disposal of mill tailings. The amount of land required is highly dependent on the grade of ore being processed. As lower grade mining and milling operations close, and are replaced by higher grade facilities, the area of land used per 200 tU/year will decrease significantly.

Releases

The releases from uranium milling include tailings, contaminated water, overburden and contaminated waste rock and airborne effluents. The tailing slurry is the most significant waste from the milling process. This waste stream is a mixture of leached solid ore and waste solutions from the grinding, leaching, uranium purification, precipitation and washing circuits of the mill. Because uranium makes up only a small part of the ore, the tailings are essentially of the same volume as are fed to mill. About 4×10^4 to $6 \times 10^4 \, m^3$ of uranium mill tailings are produced per 200 tU/year [14].

The tailings are characterised by their relatively large volumes and relatively low concentrations of long-lived natural radionuclides. About 15% of the total radioactivity originally contained in the ore is included in the yellowcake produced by the mill. Once the shorter-lived radioactive nuclides have decayed, some 70% of the radioactivity originally present in the ore is left in the tailings. The tailings contain nearly all of the naturally occurring radioactive progeny from the decay of uranium, notably thorium-230 and radium-226. The presence of thorium-230 provides a long-term source of radon.

The tailings contain any heavy metals originally present in the ore. They also include process chemicals such as ammonia and organic solvents. Therefore, if provision is not made to completely contain the material, the tailings may be a long-term source of these substances which may enter the groundwater below the impoundment. The impacts of these long-term releases must therefore be quantified and assessed. This is usually done by monitoring and predictive modelling. The topic of the environmental impacts of mill tailings has been extensively investigated and documented [14, 15].

By its nature, milling is designed to change the mineralogical and chemical characteristics of mined ore, thus putting uranium into a more soluble, concentrated form. Milling operations also tend to increase the solubility of certain contaminants associated with the ore. For example, the acid process tends to promote the dissolution and potential mobilisation of radium decay products, as well as various heavy metals present in the ore. Tailings management facilities are designed to control the release of these contaminants and/or mitigate their consequences.

Mill sites in dry areas produce little/or no liquid effluents. However any runoff water from mills sites in wet climates may contain radionuclides and may require treatment before release into watercourses. In general, contaminated water is discharged from uranium mills to tailings management facilities. The contaminants may include radionuclides, heavy metals, sulphates, chlorides, organics and ammonia. The exact mixture depends on such factors as the process technology and ore grade. Treatment of the water reduces the concentration of heavy metals, radionuclides and some anions. However monitoring of the released water is required to ensure that all constituents are within regulatory compliance. Most of the water either evaporates or is treated and discharged to the environment. In some cases the water may be recycled to the mill, thereby further reducing the amount of contaminated water.

Radioactive airborne effluents from milling may include dust and radon gas released into the air from ore stockpiles, crushing and grinding of ore, drying and packing of yellowcake, and from the tailings retention system. The amount of dust produced in the processing operations is reduced by ventilation extract scrubbers and/or filters. Tailings may be a continuing source of radon and radioactive dust after milling operation has ceased. UNSCEAR [6] estimated the release rates of Radon-222 from mill, mill tailings during operation and abandoned mill tailings to be 3 TBq, 20 TBq and 1 TBq per GWe/year, respectively.

As radon is rapidly dispersed it is difficult to detect increases over the ambient background away from the actual tailings site. Ambient concentrations are usually reached within 1 to 2 km of the site, depending on the topography [15].

Airborne chemical contaminants released to the environment include combustion products (oxides of carbon, nitrogen and sulphur) from process steam boilers, and power generation, sulphuric acid fumes in small concentrations from the leach tanks, and vaporised organic reagents from the solvent extraction ventilation system. In addition, where sulphuric acid is produced on site, sulphur dioxide is exhausted to the atmosphere, if no desulphurisation equipment is installed.

The exposure to workers in the mill may result from ore dust in the crushing and grinding areas, from short-lived radon progeny and gamma radiation where ore and tailings are handled, and from the yellowcake dust in the precipitation, drying and packing areas. UNSCEAR [6] evaluated the annual average effective dose per monitored worker to be 6.3 mSv (Table 2). Internal exposure is the greatest contribution to total exposure in milling. Results of its analysis indicate about 38% of exposure arose from the inhalation of radon progeny, about 47% from inhalation of ore dust and about 15% from external irradiation.

After termination of the operation, some form of mill decommissioning and mine and tailings closure is required by most jurisdictions. This usually includes closure of the tailings impoundment following operational, regulatory and administrative actions that would minimise long-term surveillance and maintenance. Modern tailings management facilities have been designed with permanent closure in mind. They address concerns with such things as the fluid nature of a portion of the tailings (slime), difficulties of draining, covering and revegetating the tailings mass, erosion of tailings and covers, seepages both into and out of the tailings management area, and the mobilisation of heavy metals. The latter is a particular concern where acid drainage may be released from tailings containing pyrite. Older facilities have to address these concerns at the time of close-out.

Accident considerations

The most likely types of accidents associated with uranium mill operations are inadvertent discharges of tailings to nearby streams or a major fire in the solvent extraction circuit. Inadvertent discharges may result from tailing dam failure. However, the risk of such failures may be minimised by appropriate siting, proper design, and preclosure stabilisation.

Similarly if a line carrying contaminated water or tailings should break, the impact may be mitigated by appropriate design to control and collect the spilled material.

A solvent extraction circuit contains solvent (mostly kerosene) and natural uranium, and has a serious fire potential. Conventional safety measures must be put in place to prevent or reduce the fire hazard.

From environmental hazards towards sustainable development

To meet rapidly increasing military demands for uranium, mines and mills were built rapidly from the 1940s to the 1960s, in both the west and the east. During this period, an understanding of safe working conditions and environmental protection was lacking, and priority was given to rapid production for military needs. Many problems developed because of the lack of environmental protection and radiation safety measures. In the work environment, high radon levels in underground mines caused lung cancer among workers. Mine waste and mill tailings were abandoned with no concern for environmental impacts and without taking any measures for long-term disposal.

To remediate these formerly utilised sites and resolve the associated environmental problems, projects have been undertaken in several countries to address safe disposal of the abandoned mine wastes and mill tailings. These programmes are often very costly [16, 17]. International studies and remediation projects are also currently ongoing under such organisations as the IAEA and the European Union.

By the 1970s the level of understanding of the work environment, radiation safety and environmental protection had increased in the uranium mining and milling industry. By that time regulatory agencies had been established in most uranium-producing countries. Furthermore in Australia, Canada and the USA, which are countries with large uranium resources, completing an environmental impact assessment (EIA) became a prerequisite for the start up of new mines and mills. Other countries have adopted the same requirements and today EIAs are required in many uranium-producing countries.

Government agencies have been established for monitoring releases to the air and water, as well as for land restoration following mining. In 1979, the Australian Government established the Office of Supervising Scientist in the Northern Territory to study the possible ecological impacts of uranium mining. The research started in 1979 is still continuing. Government agencies in other countries also conduct research on the ecological impacts of mining.

Most uranium producing companies have established departments responsible for activities related to safety in the work environment, including radiation safety, and for protection of the environment. Monitoring activities, as well as research and development projects are conducted to ensure the well-being of the workers and public in general and for the effective protection of the environment for many years into the future. Consulting companies have developed skills in the environmental area, and the mining companies for special projects employ such companies. The current trend and objective in many countries is to develop and operate with as few environmental impacts as can be reasonably achieved, and where practical, to restore disturbed land to its pre-mine condition.

As a result it has been found that the health and environmental effects of uranium production facilities are minimised for projects that are properly planned, developed and operated, and for which appropriate mill decommissioning and mine and tailings closure is planned and/or has been carried out. Furthermore the operators of these facilities continue to conduct research to define ways of further reducing environmental impacts.

The 1987 Brundtland report of the World Commission on Environment and Development defines Sustainable Development as "development that meets the need of the present without compromising the ability of future generations to meet their own needs".

Uranium mining and milling is moving towards sustainable development in the sense that the impact on the environment of current and future projects is being minimised. Compared with the environmental impacts of other energy sources, the effects of uranium production are relatively small. Implementation of best practice approaches in planning, operating and closing of uranium production facilities is resulting in an effective reduction of the environmental impacts. Thus future generations will be able to use large parts of decommissioned areas for some of their needs. In addition, mining and milling of uranium release relatively low amounts of carbon dioxide and other gases which cause the "greenhouse effect". Therefore, uranium mining and milling in conjunction with nuclear power have much less contribution to environmental problems such as acid rain and global warming compared to the burning of fossil fuels for generation of electricity.

From the resources point of view, uranium is widely available and should be able to satisfy future world needs even if requirements were to increases considerably. The only large scale civilian use of uranium is for nuclear fuel in the production of electricity. Known Resources (Reasonably Assured Resources and Estimated Additional Resources – Category I) of 4.35 million tU have been reported [12]. Based on geological estimates, an additional 12 million tonnes of Estimated Additional Resources – Category II and Speculative Resources can be found with continued exploration. There are also large resources consisting of low concentrations (i.e. about 100 ppm) of uranium in phosphate rock. Some uranium is already being recovered as a by-product of phosphoric acid production. More uranium could be produced from phosphates, although the cost of production would be relatively high (i.e. about \$80/kgU, or higher). There are also very large uranium resources in sea water. The recovery of such uranium has been technically demonstrated. However the cost of recovery from sea water is substantially higher than the cost of production from today's operating mines.

At present, nuclear energy production is about 2 300 TWh/year (1997). Table 4 shows the categories of estimated resources and the corresponding number of years of sustainable electricity generation assuming the present level of generation.

Present nuclear electricity production of 2 300 TWh/year corresponds to a long-term natural uranium consumption of about 50 000 tonnes per year. This takes into account some recycling of uranium and plutonium in light water reactors and a lower content of uranium-235 in depleted uranium from isotope enrichment, which will be the economic optimum at somewhat higher uranium prices than today. The estimations in Table 4 illustrate potential uranium availability to support sustainable nuclear power programmes for many generations to come.

Table 4. **Uranium resources and potential generation at current levels**

Category of resources	Estimated quantity - million tonnes uranium	Number of years of present nuclear electricity production
Uranium stocks	0.2	4
Highly enriched uranium and plutonium*	0.6	12
Known uranium resources	4.3	86
Estimated undiscovered uranium resources	12	240
Uranium in phosphates	22	440
Uranium in sea-water	4 000	80 000

* Highly enriched uranium and plutonium, resulting primarily from the conversion of nuclear weapons into nuclear fuel.

As there is no major alternative use of uranium, it may be concluded that nuclear power could be considered an appropriate sustainable development technology since it meets the present needs for energy without compromising the ability of future generations to meet their own needs. Uranium mining and milling will continue moving toward sustainable development as long as environmental management and restoration is carried out in a responsible manner.

REFERENCES

1. IAEA (1996), *Health and Environmental Aspects of Nuclear Fuel Cycle Facilities*, TECDOC-918, Vienna, Austria.

2. U.S. Atomic Energy Commission (1974), *Environmental Survey of the Uranium Fuel Cycle*, WASH 1248, Washington, United States.

3. United Nations (1979), *The Environmental Impacts of Production and Use of Energy, Part II-Nuclear Energy*, UNEP, New York, United States.

4. United Nations (1980), *Nuclear Energy and the Environment*, UNEP, New York, United States.

5. IAEA (1982), *Nuclear Power, the Environment and Man*, IAEA, Vienna, Austria.

6. United Nations (1993), *Sources and Effects of Ionizing Radiation* (Report to the General Assembly), Scientific Committee on the Effects of Atomic Radiation (UNSCEAR), New York, United States.

7. IAEA (1998), *The Impact of New Environmental and Safety Regulations on Uranium Exploration, Mining, Milling and Waste Management*, Technical Committee Meeting, 14-17 September 1998, Vienna, Austria.

   Douglas D. Chambers, Leo M. Lowe, Ronald H. Stage, *Long-Term Population Dose due to Radon-222 released from Uranium Mill Tailings*.

8. IAEA (1992), *Effects of Ionizing Radiation on Plants and Animals at levels implied by current Radiation Protection Standards*, Technical Reports Series No. 332, Vienna, Austria.

9. IAEA (1997), *Environmental Impact Assessment for Uranium Mine, Mill and In-Situ Leach Projects*, TECDOC-979, Vienna, Austria.

10. Balkau, F. and Parsons, A., *Emerging Environmental Issues for Mining in the PECC Region*, paper presented at the First Pacific Economic Co-operation Committee Minerals Forum, UNEP, 22 April 1999, Lima, Peru.

11. IAEA (1993), *Uranium Extraction Technology*, Technical Reports Series No. 359, Vienna, Austria.

12. OECD/NEA-IAEA (1998), *Uranium 1997: Resources, Production and Demand*, Paris, France.

13. IAEA (1994), *Decommissioning of Facilities for Mining and Milling of Radioactive Ores and Close-out of Residue*, Technical Reports Series No. 362, Vienna, Austria.

14. IAEA (1992), *Radioactive Waste Management*, An IAEA Source Book, Vienna, Austria.

15. Nuclear Regulatory Commission, *Characterization of Uranium Tailings Cover materials for Radon Flux Reduction*, Rep. NUREG/CR-1081, USNRC, Washington, DC (1980), in John, R.D., "Long-Term Effects of Uranium Tailings on Canadian Surface Waters", paper presented at 2nd International Conference On Radioactive Waste Management, September 1986, Winnipeg, Canada.

16. Energy Information Administration (1995), *Decommissioning of U.S. Uranium Production Facilities*, DOE/EIA-0592, Washington, DC, United States.

17. Uranerzbergbau-GMBH, *International Comparison of the Costs of Decommissioning and Restoring Uranium Production Facilities – Influencing Factors and Dependencies*, Research Contract 37/93, Federal Ministry of Economics, Bonn, Germany (Translated by OECD)].

III. ACTIVITIES OF THE INTERNATIONAL ATOMIC ENERGY AGENCY AND THE OECD NUCLEAR ENERGY AGENCY

The IAEA has assisted governments as well as uranium producers in recognising and adopting international standards, operational policies and practices that strengthen the sustainable development of uranium mining and milling. As a result, the IAEA has frequent contact with representatives of regulatory agencies and operators of uranium production facilities in countries with uranium-related activities.

The IAEA has for several years conducted related activities involving symposia, workshops, training courses, research projects, technical co-operation projects, etc. A large number of publications have been made available including safety reports and standards, safety guides, recommendations and technical reports.

During 1993 and 1994 the IAEA conducted a technical co-operation project entitled *Planning for Environmental Restoration of Radioactively Contaminated Sites in Central and Eastern Europe.* Three workshops were held in Hungary, Slovakia and the Czech Republic. The results of these workshops were published in a 3 volume, 900-page IAEA technical report (IAEA-TECDOC 865). The report gives a comprehensive review with plans for restoration of contaminated sites in Central and Eastern Europe. Many of these are related to uranium mining and milling facilities. There are also descriptions of sites in Kazakhstan and Western Europe.

The results of an IAEA project on the *Health and Environmental Aspects of Nuclear Fuel Cycle Facilities* were published in a 1996 technical report (IAEA-TECDOC–918). This report reviews important relevant issues in uranium mining and milling. Relevant information is included for Spain, India, China and the Russian Federation. A 1999 report entitled *Communications on Nuclear, Radiation, Transport and Waste Safety: A Practical Handbook,* (IAEA-TECDOC-1076), provides information related to uranium mining and milling and identifies several Internet Home Pages dealing with relevant safety and environmental aspects.

The IAEA has also launched a new initiative entitled *"UPSAT – Uranium Production Safety Assessment Team".* This is an international peer review service for uranium production facilities. This service allows the performance of safety and environmental assessments in response to invitations to the IAEA from countries.

The IAEA is currently preparing about 10 other Safety Series and Technical Reports relating to uranium mining and milling. In addition, the IAEA is planning a Symposium on the *Uranium Production Cycle and the Environment* for the year 2000 in co-operation with the NEA.

A more complete summary of recent relevant activities and publications by the IAEA is presented in Annex 2.

The NEA has actively promoted in the past the development of safe designs and methods for tailings disposal. To this effect, the NEA published in 1982 a report on *Uranium Mill tailings*

Management (OECD/NEA, ISBN92-64-02288-0) and another report in 1984 on the *Long-Term Radiological Aspects of Waste Management from Uranium Mining and Milling* [OECD/NEA, ISBN92-64-12651-1]. In 1999 the NEA is initiating a study on "Environmental Restoration of World Uranium Production Facilities". The study is in co-operation with the IAEA and under the guidance of the joint NEA/IAEA Uranium Group.

The programmes of the two Agencies have been instrumental in assisting their members in improving uranium production technology, as well as strengthening regulation of environmental and safety practices. They have also assisted in promoting information exchange to all countries in areas affecting the planning and operational stages of active production facilities, as well as in the restoration of abandoned facilities.

IV. SUMMARY OF ACTIVITIES REPORTED BY COUNTRIES

This section provides a brief summary of some of the environmental activities reported by countries and described in more detail in the corresponding national submissions. For most of the 29 reporting countries, environmental issues have become increasingly important in the last decades due to several developments affecting the uranium industry. These developments include: the large number of uranium production facilities which have been taken out of operation over the last 15 years; the increasingly stringent requirements for new facilities being imposed by many countries in the form of environmental clearance approvals; and, the restoration and reclamation measures that are being considered for many old sites which were abandoned at a time when provisions for decommissioning and rehabilitation were not sufficient.

Environmental impact assessments

In *Australia* the Environmental Protection Act 1974 provides for an environmental impact assessment process in relation to decisions made by the Commonwealth Government. The Olympic Dam Project was subject to a comprehensive environmental assessment process.

Canada has recently passed a new legislation entitled the Canadian Environmental Assessment Act. This Act replaces the "Environmental Assessment and Review Process", established in the mid-1970s. The new assessment process comprises a detailed description of each project by the mining company, a public review process by a specially appointed panel and finally a decision regarding approval by Government. Six new uranium mining projects in Saskatchewan have been, or are being, reviewed by an independent panel according to the Federal Environmental Assessment and Review Process Guidelines Order. In 1995, the environmental impact statements for the Cigar Lake and McArthur projects and the revised Midwest joint venture project were submitted for review. Public hearings for Cigar Lake and McArthur River began in September 1996. The review of the McArthur project was completed at the end of 1996 and the Panel reported to the governments in late February 1997. The recommendation was that the project be permitted to proceed subject to a number of conditions.

Emission control

Brazil is carrying out an extensive review of emissions from mining and milling. Measures have been introduced to counteract emissions to water and radon and dust emissions to air. To accomplish this, all necessary monitoring data are collected at regular intervals within a radius of 20 km around the Poços de Caldas mine and mill facility.

China monitors the content of different radionuclides in the gaseous and liquid effluents from its uranium mines and mills. This includes uranium, thorium-230, radium-226, radon-222, polonium-210 and lead-210.

The *Czech Republic* has a programme for monitoring emissions to the atmosphere and releases of mine and waste water. Emissions to the air from uranium mine and mill installations are sufficiently low that they approach the detectable limit.

At the Rössing Mine in *Namibia,* dust is a problem in the dry Namib desert environment. Several procedures are carried out to overcome this problem. Blasted rock and the haulways are sprayed with water. Only wet drilling is allowed and the ore is sprayed at the primary crusher. The cabs of all the mobile equipment are fitted with air conditioners and dust filters. Personnel working in dusty areas use respirators or "Airstream" helmets. At the tailings impoundment, which is the largest single source of dust emissions, a surface resistant to erosion of dust is maintained. Water management is also important in the dry environment. It focuses on reducing water use through economising and recycling. About 60% of water used by the mine is recycled. Seepage water from the tailings impoundment is controlled to prevent contamination of the natural ground water. Sulphur dioxide is the most important emission to the air. Several detectors are located in the area around the mine to detect release of this gas.

In *Portugal,* environmental parameters are being studied at three mines being decommissioned. Soil, sediment and vegetation samples are being collected to monitor air quality and mining effluents.

Romania uses installations and equipment such as water treatment plants for emission control and prevention of environmental contamination. Recent environmental related activities include increasing the capacity of tailings impoundments and closing ore storage facilities at various mines.

In the *Russian Federation,* the Priargunsky Mining-Chemical Production Enterprise operates a mine and mill complex 10 to 20 km from the town of Krasnokamensk in the Chiti region, Siberia. Monitoring of emissions to the atmosphere and releases to water is carried out by the company's laboratory. Radon emissions from milling tailings, low grade ores, mining waste rock dumps and open pit overburden are monitored. Completion of reconstruction of the mill tailings pond was planned for 1998. The two main environmental problems that currently exist at Priargunsky are the increasing accumulation of radioactive liquid and solid wastes, and the progressive radioactive contamination of surface and groundwater systems, which create a potential threat to the potable water supply.

Environmental aspects related to emission control at the Fe mine in *Spain* involve monitoring airborne and liquid effluents following neutralisation, depositing tailings in an impoundment, and soil stabilisation to control erosion. The implementation of the monitoring operations is carried out by specialised personnel and laboratories of the Spanish nuclear fuel company ENUSA.

In *Uzbekistan,* Mining and Metallurgical Complex is developing innovative technology for ISL uranium mining using very low concentrations of sulphuric acid in the leaching solution. This will minimise possible contamination of the ground water.

Radiation safety

Brazil has evaluated the different pathways related to the uranium mining and milling industry to determine the radiological impact of individual doses.

In Saskatchewan, *Canada*, a new technology of *Non-Entry Mining* is being developed for extracting high grade ore in underground mines. This method includes primary removal of uranium ore by boxhole boring, remote boxhole stoping or jetboring methods, all of which rely on remotely operated equipment. This technology will allow workers to stay away from the orebody and associated high levels of radiation. Another new technology proposed for use in Saskatchewan involves disposal of tailings in a paste form, consisting of higher amounts of solids (approximately 50% solids). These paste tailings are suitable for disposal under water, thereby minimising radiation exposure.

Estonia evaluated individual doses resulting from mill tailings at closed Sillamäe facility. Radon was identified as the major source of radiological impact. The annual doses are, however, relatively low at 0.02 mSv.

Gabon introduced a system for monitoring radioactivity in the environment around the mine and mill at Mounana. Several dosimeters have been installed in villages and towns close to, as well as further away from the mine. Monitoring of the ground and tap water is also being conducted.

India has installed ventilation systems to maintain radon and dust within permissible limits in its mines. All personnel working in dusty areas are provided with protective equipment such as muff respirators.

At the Rössing mine in *Namibia*, radon and thoron concentrations are regularly monitored at the mine and in surrounding areas. Various studies have been made, but no high concentrations of either radon or thoron have been detected. The average exposure of production worker at Rössing is 4.5 mSv/year, well below the new ICRP recommendation of 20 mSv/year.

Niger installed a system to monitor both the radiation doses to the employees and the radiation to the environment. This system also includes performing analysis of food stuffs such as carrots, cabbage, and tomatoes grown at different sites.

The State Geological Enterprise Kirovgeology in *Ukraine* is conducting radon monitoring of the environment around uranium mining and milling facilities. The organisation is comparing the results to international standards to determine whether a passive or active monitoring system provides better results.

Decommissioning and restoration

Argentina is preparing for the decommissioning and closure of the Malargüe mine and mill in the Mendoza Province. This is a long procedure including planning, permitting, and implementing the closure, as well as carrying out verification for a period of 20 years.

In *Australia*, recognising and establishing close-out criteria for uranium mines is a high research priority. This includes the provision of standards against which rehabilitated mines may be assessed to determine their acceptability or otherwise, rehabilitation measures specifying the required types and periods for monitoring facilities. Another priority is ensuring the environmental integrity of final tailing disposal techniques.

Bulgaria decided to close all uranium mines as they were found to be uneconomic under today's market conditions. A plan to liquidate and restore the mining and milling sites is being implemented. As part of this activity 128 studies, including radioecological and hydrological surveys, were conducted by expert teams.

Canada has a new *Policy Framework for Radioactive Waste Disposal*. Waste producers and owners are responsible, in accordance with the "polluter pays" principle, for the funding, organisation, management and operation of disposal and other facilities required for disposal of their wastes.

Pursuant to the *Atomic Energy Control Act*, uranium producers are required to provide financial assurances at the earliest stages of operations for the eventual decommissioning of uranium production facilities and sites. More than a dozen means of providing the necessary financial assurances have been identified. In June 1996 an Environmental Assessment Panel submitted its recommendations to the Minister of Environment for decommissioning of the Elliot Lake uranium tailings. The recommendations are in agreement with the proposals made by Rio Algom Limited and Denison Mines Limited. The Federal response, released in April 1997, approved most of the Panel's recommendations.

China has been decommissioning 6 uranium mines and 3 mills since 1986. Most of the decommissioning projects mainly involve the treatment of low grade waste ore and tailings. In 1993, the government published the "Technical Regulations for Environmental Management and Decommissioning of Uranium Mining and Milling Facilities".

In the *Czech Republic*, underground water near the Stráz deposit in Northern Bohemia has been affected by the intensive application of ISL mining for more than 20 years. Now two aquifers with a total area of 32 km^2 are contaminated. A remediation programme was accepted by the Czech government in 1996. The planned remediation involves extracting the soluble salts from the contaminated aquifers. The remediation target following ISL extraction is to gradually:

- Lower the content of dissolved solids in the Cenomanian aquifer to a level that will reduce the risk of endangering the Turonian aquifer – an important resource of potable water.

- Lower the content of dissolved solids in the Turonian aquifer to the water quality limits of Czech drinking water standards.

- Re-establish the surface ecosystems over the leaching fields.

The Czech company DIAMO is at present carrying out a large number of remediation and decommissioning activities including:

- Decommissioning of the Hamr, Olsí, Jasenice-Pucov, Licomerice-Brezinka, Vítkov II and Zadní-Chodov mines.

- Decommissioning of the Stráz chemical processing plant.

- Remediation of the tailings impoundments at Stráz, Bytíz-Príbram, and MAPE Mydlovary.

- Processing of impoundment water at Príbram and mine water at Zadní-Chodov and Okrouhlá Radoun.

- Recultivation of the Jeroným Abertamy waste dump in the Jáchymov region.

- Construction of a mine water processing plant at Horní Slavkov.

Finland conducted uranium mining on a small scale at Lakeakallio and Paukkanvaara in the 1957-1965 period. The pits and waste heaps at these sites have been covered by soil. The radiological state of the former mines have been monitored by the Finnish Centre for Radiation and Nuclear Safety since 1974.

In *France*, Le Bernardan-Jouac mine is operating, while remediation work is ongoing at the Lodève, St Pierre du Cantal, Bertholène, La Crouzille Bessines and L'Écarpière mines. Remediation is complete for the St Hippolyte, Lachaux, Rophin, Gueugnon, Bois Noirs St Priest, Le Cellier, La Ribière and La Commanderie mines. Radiological monitoring is continuing for the last group. In 1994,

Cogema spent 129 million French francs on remediation work at mine sites. The mine remediation policy at Cogema is to fulfil all laws and regulations as set out by the company's Department of Safety, Quality and Environment. Cogema's Environmental Service of Mine Sites of the Uranium Branch performs the work. This work is inspected by the DRIRE authorities (Regional Authorities for Research, Industry and Environment). The type of remediation work required at each site depends on the type of mining activity carried out and the local conditions.

For each site, an extensive study is made first before the implementation of any restoration.

Agreement is then reached with the stakeholders, such as local inhabitants, regarding the acceptable closure procedure. Open pit mines may then be rehabilitated into such features as water reservoirs for agriculture, scuba diving training centres or fishing lakes depending on the local requirements. Such reclamation works are preceded by investigations to ensure the radiation and general site safety. At underground mines, access is first effectively stopped to prohibit accidents. As the water level increases in the mine following termination of pumping, mine effluents are monitored and controlled to prevent any damage to the environment. The most important remediation activity concerns the disposal of mill tailings.

In *Germany*, a total of 216 000 tU were produced from 1946 to 1990 by the former Soviet-German Company SDAG Wismut. Following the closure of commercial uranium production in 1990, the total area of mining and milling sites held by Wismut was 37 km^2 and the total affected area was 240 km^2. In 1991, a programme was initiated by the German Government to evaluate the extent to which clean-up activities would be necessary.

Wismut GmbH is responsible for the decommissioning and remediation of its sites and facilities. License applications have to be submitted to the relevant state authorities (Saxony or Thuringia). The total cost for decommissioning and rehabilitation of all Wismut facilities is estimated at 13 000 million DM and this will take about 15 years.

Major selected rehabilitation projects include:

- Open pit mining produced about 600 million m^3 of ore and waste rock. About 160 million m^3 of this amount is from the Lichtenberg open pit. About 80 million m^3 were backfilled before 1990. About 100 million m^3 of waste material remains in dumps located around the town of Ronneburg. Most of the material is to be backfilled in the open pit.

- Underground mining yielded about 300 million m^3 of material, of which half was ore. In the Aue district, there are about 40 piles with a total volume of 45 million m^3 covering an area of about 3 km^2. A major programme to stabilise, reshape, cover and revegetate these piles is ongoing.

- Tailings were produced at 2 large conventional mills operating at Crossen and Seelingstädt. At Crossen, alkaline leaching was used to process ore from the Erzgebirge. The tailings were disposed of in a nearby pond. It is about 2 km^2 in size, contains about 45 million m^3 of tailings and 6 million m^3 of water. Seelingstädt used both alkaline and acid leaching to process ores mainly from the Ronneburg district. The tailings were disposed of in 2 nearby ponds covering 3.4 km^2, with a volume of 107 million m^3.

Maximum exposure limits have been established for the tailings following reclamation. Radiation doses to the general public resulting from the former operations should be less than 1 mSv/year. Additional detailed criteria have also been established.

In *Hungary*, the government decided to close the Mecsek Mine at the end of 1997. A conceptual plan has been developed defining the methodology of the remediation work for the 65 km^2 area. It includes soil investigation, drainage, covering of leach piles and tailing ponds, as well as community planning and many other details. The closure activities, remediation and community planning must be finished by 2002. The expected cost is USD 80 to 85 million.

In *Japan*, wastes from past uranium production have been monitored and controlled in accordance with all related regulations.

In *Kazakhstan*, there are 15 uranium deposits that have been mined out, shut down or conserved for the future. The sandstone hosted uranium deposits occur in sedimentary basins that also host large volumes of groundwater. The contamination of groundwater related to the uranium deposits, both occurring naturally and resulting from ISL mining, led to the development of an exclusion zone equal in size to 150 x 150 km. The extraction of drinking water from this zone is prohibited. Forty years of uranium production by conventional mining resulted in an accumulation of about 200 million tonnes of waste dumps and mill tailings. Systematic investigations of radioacitve mine wastes started in the 1990s. In 1996, within the framework of the European Union's support programme to CIS countries (TACIS), the first stage of an "Assessment of Urgent Measures to be taken for Remediation of Uranium Mining and Milling in the Commonwealth of Independent States" was carried out. A catalogue of 100 sites with radwaste was compiled. From these, 5 sites were chosen for further study. The influence on the nearby community is to be investigated. Based on the results of stages 3 and 4 of the assessment, a restoration and rehabilitation programme will be formulated for Kazakhstan.

In the *Russian Federation*, at the Lermontovskoye State Company "Almaz", restoration work is underway at two depleted underground uranium mines: Beshtau (closed in 1975) and Byk (closed in 1990). Beshtau and Byk produced 5 700 tU. From 1965 to 1989 ore was processed not only by traditional sulphuric acid leaching, but also using in-place and heap leaching. From the 1980s until 1991, ore from the Vatutinskoye deposit, Ukraine, and the Melovoye deposit, Kazakhstan was also, processed at the Lermontovstoye mill, after which uranium production was stopped. Since that time the plant has been used to process other types of ore. The company's laboratory is now monitoring the emissions from the contaminated area. The air is monitored for emissions of radon, polonium-210 and radium isotopes, and the surface water for uranium and thorium nuclides, radium-226 and polonium-210. Decommissioning and restoration of the mine and mill site is in progress. Rehabilitation of the 36 hectare mine waste dumps at Beshtau is nearly complete. Rehabilitation of the 18 hectare waste rock dumps at Byk is underway and is planned to be completed in 1999.

Rehabilitation and decommissioning of the 118 hectare Lermontovstoye mill tailings started in 1996 and is planned to be completed in 2005. In 1996 a radiation survey of 3 200 hectares in the region was conducted. The remediation of the mill buildings and area is being planned. Rehabilitation activities are reimbursed from the state budget and are conducted according to state instructions.

Slovenia established a decommissioning plan for the Zirovsky vrh uranium production centre, including permanent protection of the biosphere against the consequences from uranium mining.

In *Spain,* regulations require the owner of a facility to present a *decommissioning plan* to the Ministry of Industry and Energy prior to finalising productive activities. This decommissioning plan must include detailed information about criteria for decommissioning, analysis of the radiological impact on the public and a monitoring programme. The Lobo-G plant, located in Badajoz, has been closed and restored. The project included an open pit mine, a mill and a tailings pond.

The uranium processing plant in Andujar, located in Jaén, including a mill and tailings ponds, has been closed. In 1995, a ten-year supervision programme was initiated at the Andujar uranium concentrate plant following the completion, in 1994, of the dismantlement and restoration. Several small abandoned uranium mines where mining was carried out in the 1950s or 1960s are also included in the General Remediation Plan. Decommissioning of the waste dump and processing plant at ENUSA La Haba Centre, Badajoz province, was started and scheduled to be completed in early 1997. At the Fe operating mine in Salamanca, two projects for decommissioning have been presented to the Nuclear Regulatory Council. The first is the old Elefante treatment plant and the second is the heap leaching operation. Approval has been given by the authorities for restoration of 19 old uranium mines which operated between the 1950s and 1981.

In *Sweden*, the uranium mine at Ranstad was restored in the early 1990s. The open pit has been transformed into a lake to be used for bathing and canoeing. The tailings have been covered with a 1.6 metre thick protective layer consisting of moraine and other materials. Some of these are permeable to air. This is a concern because of the possible reaction between oxygen and sulphur in tailings, resulting in production of sulphuric acid which dissolves heavy metals into waste water. Ranstad is in the supervision phase following restoration.

Ukraine has initiated programmes to rehabilitate and recultivate several underground and open pit mines.

The uranium production programme of the *United States* has left a large number of mines and mills, many of which required decommissioning, close-out and/or restoration. In the US the regulation and oversight of uranium mine and mill sites is the responsibility of different government organisations. There are a total of 26 uranium mills involved in production of uranium for commercial purposes. One is in operation, 5 are on standby and 20 are at various stages of decommissioning. In addition there are 24 "inactive" uranium mill sites that produced uranium for the government programme prior to 1970. The cost of remedial action for these 24 sites (and associated vicinity properties) are born by the US government under terms of the Uranium Mill Tailings Remedial Action Programme (UMTRAC). The responsibility for the decommissioning and restoration of the other mills depends on the purpose for which the uranium was produced. Those mills that produced uranium for commercial purposes only are the responsibility of the owners and operators. Of the 20 mills undergoing decommissioning, for those that produced uranium for both the government and civilian programmes and have commingled tails, the cost is distributed between the US government and the operator. The Government proportion of remedial action costs for the 13 uranium mill sites (commingled) out of the 20 sites are reimbursable by the Government under Title X programme. The report *Decommissioning of US Uranium Production Facilities,* (DOE-ERA 0592, February 1995), examines the cost of decommissioning 25 conventional mills and 17 non-conventional sites in the USA.

The US Nuclear Regulatory Commission conducts regulation of uranium mills in the United States. However, its authority does not extend to uranium mines. The closure and restoration of mine sites is carried out according to regulations of the individual States by treating uranium mines like any other type of mine. Most states have laws on abandoned mines and regulate reclamation work. Mines located on federal land are subject to federal land management laws and regulation which also govern mined land reclamation.

Regulations

In *Australia* uranium mining is regulated both by Commonwealth, Northern Territory and South Australia State Legislation. The main Commonwealth Acts are: Atomic Energy Act 1953 as amended, Environmental Protection (Impact of Proposals) Act 1974 and Environmental Protection (Nuclear Codes) Act 1978.

In *Finland*, mining is regulated by the Mining Law of 1965 or by the Land Extraction Act of 1987. The Nuclear Energy Act of 1987 regulates uranium activities. These laws include regulations concerning the opening, operating and shut-down of uranium mines. New considerations result from the Environmental Permit Procedure Act of 1992.

In *France*, minerals occurring underground belong to the State. The State may however, transfer the exploitation of such deposits to mining companies. There are the general mining laws of 1955 and 1994 which describe the procedure for closing a mine. The regional authorities monitor mine closure and assure that it is performed in accordance with the prescribed regulations. Mining and mine restoration must be conducted in accordance with the Environmental Laws of 1977 and 1996.

In *Germany*, the restoration of the Wismut mines and mills is regulated by the Federal Mining Law, The Atomic Energy Act, The Federal Emission Protection Law, The Environmental Liability Act and The Radiological Protection Ordinance. The Legislation for Radiation Protection was established in the former German Democratic Republic. This legislation was recognised in the agreement for the reunification of Germany and is still in effect for uranium mine rehabilitation in eastern Germany.

In *Hungary,* uranium mining is regulated by Mining Law No. XLVIII/1993, and Governmental Regulation No. 152/1995 on the Environment as well as the Act on Atomic Energy 1996. In addition, regulations are based on IAEA Technical Report No. 362 *"Decommissioning of Facilities for Mining and Milling of Radioactive Ores and Close-out of Residues"* and the 1991 Report of the National Radiobiological and Health Institute.

In *Japan* the relevant laws include: Mines Safety Act, Mine Safety Regulations, Nuclear Raw Materials, Nuclear Fuel and Nuclear Reactor Act and Regulation for Nuclear Raw Materials and Nuclear Fuel Materials Milling Operation. The Mines Safety Act requires the responsible party to prevent mine pollution from waste rock and tailings. The Atomic Control Act applies to milling and requires that safety measures be taken when disposing of tailings to avoid environmental contamination.

In *Kazakhstan* there are two relevant laws regulating uranium activities: the law on the *Use of Atomic Energy* and the draft law on *Radwaste Management*.

In *Spain,* Law No. 25/1964 on Nuclear Energy specifies all aspects of uranium mining. Law No. 15/1980 for the Setting up of the Nuclear Safety Council specifies the responsibilities of the authority. Decree No. 2869/1972 on Nuclear and Radioactive Nuclear Facility Regulations classifies uranium ore processing plants as first category radioactive installations. Decree No. 53/1992 on Health Protection Regulation Concerning Ionising Radiation establishes the basic standards for radiation protection for both workers and the general public. EURATOM directives No. 80/836 and No. 84/467 were incorporated in Spanish legislation through Decree No. 53/1992.

In *Ukraine*, the Law relating to Mining and Treatment of Uranium Ore includes regulations on mining and milling, public and environmental safety, and radioactive methods and procedures.

In the *United States*, the 1978 Public Law No. 95-804 Uranium Mill Tailings Radiation Control Act (UMTRCA) regulates the clean-up and stabilisation of uranium mill tailings at inactive uranium processing sites. The applicable environmental standards and guidelines for this law are established by the US Environmental Protection Agency (EPA) and are contained in US Federal Law Health and Environmental Protection Standards for Uranium and Uranium Mill Tailings (40CFR Part 192). Title X of the Energy Policy Act, 1992 establishes the authority and framework for providing Federal assistance for decommissioning and clean-up of sites that formerly produced uranium for both the civilian and government programmes.

The closure and reclamation of uranium mines is in most states regulated by laws concerning abandoned mines and reclamation work. Reclamation of uranium mines located on Federal land are subject to rules of the US Department of Interior's Office of Surface Mining Reclamation and Enforcement, or by the US Bureau of Land Management under The Federal Land Policy and Management Act of 1976 (Public Law No. 94-579).

Miscellaneous

In *Finland*, an international research programme on radionuclide transport analogy has been started in the surroundings of a uranium deposit at Palmottu in southwestern Finland. The deposit is situated within a former prospection target area. The cored drill holes still remain open and offer suitable conditions for hydrogeological and geochemical studies. The programmes main objective is to study the environmental effects of a uranium deposit in its natural state.

In 1997, *Kazakhstan* started the Technical Co-operation Project Modern Technologies for In-Situ Leach Uranium Mining in co-operation with IAEA. The project objective is to minimise the health and environmental impacts of ISL uranium mining, with emphasis on controlling or eliminating contamination of groundwater aquifers that host uranium ore. Lessons learned in the project will be useful for licensing ISL uranium production operations.

V. NATIONAL REPORTS

This part of the report presents the national submissions on environmental activities in uranium mining and milling provided by the official government organisations or the members of the Joint NEA/IAEA Uranium Group (Annex 3). It is important to note that the national reports do not necessarily include all the ongoing environmental activities or issues in each reporting country. Furthermore, it should be noted that the national boundaries depicted on the maps are for illustrative purposes and do not necessarily represent the official boundaries recognised by the Member countries of the OECD or the Member states of the IAEA.

• Argentina •

ENGINEERING PLAN FOR THE RESTORATION OF THE MALARGÜE INDUSTRIAL COMPLEX

Background

The CNEA (Comisión Nacional de Energía Atómica) operated the Malargüe Industrial Complex, at Malargüe, in the Mendoza province for 32 years. The plant produced yellow cake from uranium ores mined from several deposits in the province. The production was stopped at the end of 1986, and site restoration was started. Restoration for the 700 000 t of tailings is to be undertaken according to a schedule that includes research studies used to establish the best management plan.

In 1993, the document on "Environmental Impact Assessment and Long-Term Management of Uranium Tailings from the Fabril Malargüe Industrial Complex" included the decision to dispose the wastes by relocating them to another site within the Malargüe complex. The detailed engineering plan was defined to comply with the design requirements and proposed objectives, according to the basic engineering criteria and standards.

The objective of the document is to establish the engineering plan for the management of the uranium mill tailings within the decommissioning plan for the installation. The final objective is to stabilise the tailings system for the long term and to minimise the release of pollutants into the atmosphere to the lowest levels that may be reasonably achieved.

The document is divided into chapters. Each chapter includes analyses of topics comprising the engineering which are technically supported by additional studies included in Annexes and Attachments. The necessary procedures and documents for the execution of the corresponding works may be developed from the information available.

Area of arrangement

The Malargüe Industrial Complex includes various facilities and the current tailings arrangement. To prepare the new site for the wastes, several tasks must be completed such as the dismantling and demolition of existing installations and equipment, decontamination of soils and the preparation of the foundation floor. The current topography of the site was mapped and the total amounts of materials for disposal (process tailings, contaminated soils and wastes from demolition) were assessed. At the selected site, open pit wells were dug in order to determine the characteristics of the foundation soil. Based on the results obtained from the analysis and evaluation of the soil samples, the first actions required to construct the foundation are: excavation of the first thirty centimetres, scarifying and compacting this into a layer in order to homogenise the base and avoid differential settling.

Engineered barrier

The proposed containment system is composed of an engineered, layered barrier of natural materials. Compaction of the soil is followed by placement of the lower barrier of the system. This barrier consists of a porous compacted layer, a compacted sand-slime soil layer, and a compacted clay layer.

Tailings management

The mechanical characteristics of the tailings were investigated through lab and field tests. The tests were carried out by the Soils Laboratory of the Geotechnic Department of the Universidad National de Cuyo (UNC). The saturation level and the withholding capacity of the residues were also assessed. The summary of these data allowed the selection of the best methodology to remove the material from the lagoon.

Furthermore, vertical extraction of the tailings was decided, according to the granulometric studies and pilot assays on terraces of work scale bank. The tailings would be moved to the new site by trucks. Water is added to the tailings to reduce dust pollution. Wastes would be arranged in compacted layers, and neutralised with lime. As the materials are heterogeneous, they would be located according to their mechanical behaviour, i.e. those of greater strength would be located in the lower part of the system and those of lower strength in the higher part. The physical and chemical features of the residues were studied. They are noted in the Environmental Impact Assessment document.

Stabilisation and coverage

The objectives of the multilayer barrier are to reduce radon release and gamma radiation, to minimise the infiltration of rainfall, to prevent the dehydration of the clay layer and to provide a long-term barrier against erosion. The multilayer barrier is composed of a compacted clay layer and a compacted sand-slime soil layer, covered by a rock layer.

Structural design

The structural design of the containment system is accomplished by taking into account the characteristics of the different types of soils that will be contained and the improvement of the physical and chemical conditions, aiming at achieving the long-term stabilisation of the residues.

A field assay was carried out consisting of a test bank of the more representative materials. The objective was to analyse the equipment behaviour, materials response and the most adequate methodology for the supply of lime, and to determine the time required for drying.

Hydrology and hydrogeology

Taking into account the hydrogeological studies carried out and specially the need of stabilising the system for a very long term, the construction of underground drainage is recommended in order to lower the underground water level. In this way an extra barrier is added between the wastes and the water table, in the event of any abnormal increase in the amount of underground water.

To allow free drainage of the rainfall, and considering the contribution of the wastes southwest flank and the increase in the water table, superficial drainage is proposed to be located over the tables that coincides with the greater elongation of underground drainage.

Monitoring period and rehabilitation of the area

A 20-year monitoring period is planned. The objective is to evaluate the performance of the tailings system in relation to the model used. A monitoring programme will be implemented in order to detect possible modifications of the expected conditions. In the event that such anomalies should occur, the corresponding corrective actions shall be taken.

The last part of this remediation programme will be the reforestation of the area. This is done to make the area available for public use, keeping in mind the restrictions and final approval of "Ente Regulador Nuclear" (Nuclear Regulatory Agency).

• Australia •

URANIUM MINING IN AUSTRALIA – TODAY

Australia is estimated to have approximately 26% of the world's resources in the low cost uranium reasonably assured resources (RAR) category, although there are currently only two operating uranium mines on the continent. The Ranger mine is located in the wet/dry tropics in the World Heritage listed Alligator Rivers Region of the Northern Territory, which also contains the Kakadu National Park. The Olympic Dam uranium/copper/gold mine is in an arid region of South Australia, and lies on the margin of the Great Artesian Basin, from which water for the project is drawn. Ore extracted in 1979 at Nabarlek, which is also in the Alligator Rivers Region, was milled until 1988. Rehabilitation of the pit (into which the tails were deposited), evaporation ponds, and the mill were completed in 1995. Environmental monitoring of the Nabarlek site continues.

Australian Commonwealth (Federal) Government policy limited uranium mining activities to these three mines (the "Three Mines Policy") from 1983 to 1996. A change of government in 1996 and the abolition of the Three Mines Policy, has renewed interest in other uranium orebodies which have not yet been exploited.

There are many other identified Australian uranium orebodies in addition to those listed in the table. Australia's total low cost RAR recoverable at costs of less than USD 80/kgU, as at December 1996 was 622 kt U_3O_8.

Operating and formally proposed Australian uranium mines

Mine/Deposit	Proved U_3O_8 Resources	Status
Ranger No. 3	53.4 kt @ 0.28%	Production rate 5 kt U_3O_8 per year.
Olympic Dam	360 kt @ 0.06-0.08%	Producing 1.6 kt/year U_3O_8, planned increase to 4.6 kt/year from late 1999, environmental impact statement for Olympic Dam Expansion released for public comment in May 1997.
Jabiluka	90.4 kt @ 0.46%	Final Environmental Impact Statement released in June 1997. Government approval in October 1997.
Kintyre	24 kt @0.2-0.4%	Guidelines for Environmental Impact Statement finalised.
Beverley	12 kt @ 0.27%	Environmental Impact Statement in preparation.

Regulation of uranium mining in Australia

As Australia consists of States and Territories federated under the Australian Commonwealth Government, each State and Territory (apart from the Australian Capital Territory, which does not allow mining) has independently enacted legislation to regulate the mining industry, and has established government agencies or departments to administer and enforce that legislation. It is in the States and Territories where the primary responsibility and power is vested with respect to the regulation of the environmental aspects of uranium mining and milling activities.

The Australian Commonwealth (Federal) Government does not participate in the day to day regulation of uranium mines. It has in place more general Commonwealth legislation which ensures that Australia meets its international responsibilities, including those promulgated in treaties and conventions to which Australia is a signatory. Other than in the Alligator Rivers region of the Northern Territory, Australian Commonwealth Government involvement in regard to environmental issues associated with uranium mining and milling ends at the completion of an environmental impact assessment (EIA) process required by the *Environment Protection (Impact of Proposals) Act 1974*. This Commonwealth Act provides that a) the proponent of an environmentally significant development be required to prepare an Environmental Impact Statement (EIS) for public comment, and b) that the Australian Commonwealth Government assesses the EIS.

The environmental impact assessment process provides a vehicle for the Australian Commonwealth Government to review development proposals, such as uranium mines, and veto those developments which are found to have unacceptable environmental impacts. At the conclusion of the EIA process, the State or Territory government in which the development is to be located assumes full responsibility for the environmental regulation of the uranium mine. To avoid duplication, the EIA of new uranium mines is usually conducted jointly by the Commonwealth and the relevant State or Territory.

SUMMARY OF LEGISLATIVE AND REGULATORY FRAMEWORK

General

The *Environmental Protection (Impact of Proposals) Act 1974* provides for an environmental impact assessment process in relation to decisions made by the Commonwealth Government.

With the exception of the Northern Territory (NT), uranium mining operations in Australia are regulated by State Governments, which own the mineral resources. In the NT, the Commonwealth owns the uranium, resulting in NT specific regulatory and legislative arrangements which are discussed in more detail below. In all Australian States, except Victoria and New South Wales, uranium mines are developed under the relevant mining/resources development acts and the regulation of these mine sites is undertaken by State Government authorities under relevant State legislation. The States are also responsible for occupational health and safety standards relating to radiation exposure at these mines.

Victoria and New South Wales have enacted legislation to prohibit prospecting for, and mining of uranium.

The Commonwealth Government is responsible for administering export controls on uranium under the Customs (Prohibited Exports) Regulations. The current Government's policy is to approve the mining and export of uranium from any project provided stringent environmental, heritage and nuclear safeguards obligations are satisfied.

Northern Territory – Alligator Rivers Region

In the Alligator Rivers Region, the Commonwealth Government plays a more prominent role in the protection of the environment from the potential impacts of uranium mining. The Alligator Rivers Region includes the Kakadu National Park, which is World Heritage listed for its cultural, landscape and ecosystem values. The Alligator Rivers Region is Australia's foremost uranium province. Three of the principal deposits lie in leases surrounded by the national park, as they were granted before declaration of the park and are excluded from it. They are Ranger, which commenced in 1980 and has a life extending to about 2012; Jabiluka, for which development approval has been given, and Koongarra.

The Nabarlek uranium lies close outside the Kakadu National Park area. Production from the milling of stockpiled ore ceased in 1988 and the site was successfully rehabilitated in 1994-95. Exploration is in progress in areas outside the national park, but no mining exploration activity is permitted within the park (outside of the three existing mining lease exclusions).

In 1977, the Australian Commonwealth Government completed a major public inquiry to determine whether to allow uranium mining in the Alligator Rivers Region, at a time when the outstanding environmental and cultural values of the region were coming to public notice. The government accepted the inquiry's recommendation that uranium mining be permitted subject to a suite of conditions, one of which was the creation of the position of Supervising Scientist to oversee the environmental aspects of uranium mining operations in the region. Through the Supervising Scientist and the organisations which report to him (the Office of the Supervising Scientist – now Uranium Mining Audit and Review Branch – and the Environmental Research Institute of the Supervising Scientist), the Australian Commonwealth Government facilitates the protection of the Kakadu National Park from the potential impacts of uranium mining in the region.

Legislation limits the Supervising Scientist Group's activities to the Alligator Rivers Region, where it fulfils an oversight role; the regulatory role is the responsibility of the Northern Territory Government. An important aspect of the Supervising Scientist's approach is to measure environmental performance at the mines through a twice-yearly audit process, the results of which are presented to a meeting of government, industry and community representatives, where environmental performance information, environmental reports, monitoring data and stakeholder concerns are discussed. The Supervising Scientist has consistently reported that no significant environmental impact on the Kakadu National Park has arisen from the Ranger uranium mine.

Uranium mining in the Northern Territory is regulated by both Commonwealth and NT legislation (DPIE, 1996). There is a range of Commonwealth and Territory Acts which regulate mining operations in general, however, only the main Acts which regulate uranium mining are summarised below.

Commonwealth legislation

Atomic Energy Act 1953

The Atomic Energy Act gives effect to the Commonwealth's retention of ownership rights to uranium in the NT. Consistent with the Commonwealth's ownership of uranium, all royalties from uranium mining in the NT are set by, and payable to the Commonwealth.

The Atomic Energy Act also provides the basic authorisation for mining activities in the Ranger Project Area, which are being conducted by Energy Resources of Australia (ERA). However, because the area is Aboriginal Land granted under the Aboriginal Land Rights (Northern Territory) Act, all royalties received by the Government from Ranger are paid into the Aboriginal Benefits Trust Account.

The second report of the Ranger Uranium Environmental Inquiry (Fox Inquiry), released in 1977, made recommendations about specific requirements for uranium mining and milling which were aimed at protecting the environment in the Alligator Rivers Region from the effects of uranium mining. These environmental requirements were developed by the Commonwealth based on the Fox Inquiry's recommendations and form part of the regulatory and contractual arrangements applying to the Ranger project. For example, they are included in the authorisation issued under the Atomic Energy Act for mining activities in the Ranger Project Area.

Environmental protection (impact of proposals) Act 1974 (EPIP)

This Act provides for an environmental impact assessment process in relation to decisions made by the Commonwealth Government. The Act ensures, to the greatest extent that is practicable, that matters affecting the environment to a significant extent are fully examined and taken into account in the making of decisions and recommendations by, or on behalf of, the Australian Government. As the Commonwealth must approve any uranium exports, all uranium mining projects must comply with this Act.

Environmental protection (Alligator Rivers Region) Act 1978

This act provides for the appointment or establishment of:

- A supervising scientist for the Alligator Rivers Region (as part of the Supervising Scientist Group – SSG).

- A co-ordinating committee for the Alligator Rivers Region.

- An Alligator Rivers Region research institute (now Environmental Research Institute of the Supervising Scientist – ERISS).

The main functions of these authorities and organisations is to co-ordinate and supervise measures for the protection and restoration of the environment of the Alligator Rivers Region from the effects of uranium mining. The task of the SSG is to collect base data and establish environmental standards, plan and supervise an environmental programme, and undertake research programmes.

Environment Protection (Nuclear Codes) Act 1978

Under this act, three codes of practice in relation to radioactive ores and wastes have been published with two of these being revised later. The three codes, with the latest dates of publication are:

- Code of Practice on the Management of Radioactive Wastes from the Mining and Milling of Radioactive Ores, 1982 (Waste Management Code).

- Code of Practice on Radiation Protection in the Mining and Milling of Radioactive Ores, 1987 (Health Code).

- Code of Practice for the Safe Transport of Radioactive Substances, 1990 (Transport Code).

These three codes of practice are consistent with internationally agreed standards, and therefore provide a sound basis for radiation protection in the mining and milling of radioactive ores and the transport of radioactive materials. Whilst these are Commonwealth instruments, the State/Territory enforces them. The codes have been adopted as regulations in the NT.

Northern Territory legislation

The *Mining Act 1982* provides for the grant of a mineral lease for the purpose of mining. All mineral leases are granted subject to the condition that the holder will carry out its mining programmes and other activities so as to cause as little disturbance as practical to the environment, and comply with written directions from the NT Department of Mines and Energy.

The *Mine Management Act 1990* provides for the inspection and management of mining operations.

The *Uranium Mining (Environment Control) Act 1979* aims at providing for the environmental regulation of all uranium mines in the NT, including those in the Alligator Rivers Region. The Act reflects the agreed Commonwealth/NT approach for mining operations to be regulated as far as possible through NT law.

Given the overlap of NT and Commonwealth jurisdiction in relation to environmental regulations in the uranium mining industry, the Commonwealth and NT Governments have developed working arrangements to delineate the roles of the various agencies involved and establish appropriate consultative arrangements.

Mine site rehabilitation

The Ranger mine is subject to a unique set of arrangements governing its eventual rehabilitation. The Fox Inquiry recommended that arrangements be made to ensure that the operator carried out required rehabilitation work, and that the costs of this work be paid by the operator. This was done via the Ranger Agreement (between the Commonwealth and the mining company, ERA) which contains a number of detailed provisions to ensure that the Ranger Project Area is satisfactorily rehabilitated by the company.

Under the agreement, ERA is required to prepare an annual rehabilitation plan on the basis of the cessation of mining by the 31st of March. Once this plan is accepted by the Minister for Resources and Energy, it is independently costed and adequate security to cover the cost of implementing the plan is provided by the company to the Ranger Rehabilitation Trust Fund.

The Nabarlek project was developed under an NT mining lease, and hence the rehabilitation requirements were determined by NT legislation. The company recently completed the dismantling and decommissioning programme on the site and the former mine/tailings pit and evaporation ponds have been revegetated. The site will, however, be subject to a longer term monitoring programme.

South Australia – Olympic Dam operations

The legal framework for the terms and conditions of project development and operations at Olympic Dam are set out in the indenture between the South Australian Government (the State) and WMC Ltd. This was ratified by the State Parliament through the Roxby Downs (Indenture Ratification) Act 1982 (the Indenture). The mine is primarily a copper mine, but also produces uranium and gold. The Indenture, which was amended in 1996 to provide for the development of a project recovering 350 000 t/a of copper metal (formerly 150 000 t/a), and associated products, defines the obligations of the State and WMC Ltd in relation to the project.

The initial Olympic Dam Project and the recently completed Olympic Dam Expansion Project were both subject to a comprehensive environmental assessment process. The draft Environmental Impact Statement (EIS) and the Supplement together with the State Assessment Report comprise the

Final EIS. The Commonwealth Government also assessed the adequacy of the EIS in terms of the Administrative Procedures under the *Environment Protection (Impact of Proposals) Act 1974*.

The obligations of the company and the State are set out in the Roxby Downs (Indenture Ratification) Act 1982 as amended. As outlined in the Indenture, WMC Ltd must comply with relevant State and Commonwealth legislation and Codes of Practice relevant to environmental issues.

In regard to ongoing environmental management, the Indenture Agreement requires that a programme for the protection, management and rehabilitation of the environment, including arrangements for monitoring areas in order to ascertain the effectiveness of such a programme, shall be submitted to the Minister every three years.

The main Commonwealth Acts which regulate the mining operations are as above. A complete list of all the State and Commonwealth Acts which relate to uranium mining at Olympic Dam are given in WMC Ltd. (1996); they include the following:

- Controlled Substances Act 1984.

- Dangerous Substances Act 1979 and regulations.

- Environment Protection Act 1993 and regulations.

- Environment Protection Act (Industrial Noise) Policy 1994.

- Environment Protection Act (Air Quality) Policy 1994.

- Mining Act 1971 and regulations.

- Public and Environmental Health Act 1987.

The *Radiation Protection and Control Act 1982* regulates activities relating to uranium mining, processing, and tailings disposal by calling up the relevant Commonwealth Codes of Practice, i.e. the Health Code and the Waste Management Code. This Act also regulates all transport of radioactive goods, including uranium oxide concentrate for export, by calling up the Transport Code.

The Occupational Health and Safety and Welfare Act 1986 and consolidated regulations sets out the general duty of care requirement, and implicitly requires the formulation of an Occupational Health and Safety Management System. This legislation draws on the Health Code, Waste Management Code and Transport Code.

Mine site rehabilitation

The State Government is responsible for regulation of mine site rehabilitation. The original EIS for Olympic Dam and the EIS for the Expansion Project both describe a conceptual design for the eventual rehabilitation of the site, particularly the tailings retention system. This concept is subject to ongoing development via experience gained from the progressive rehabilitation of sites within the project area that are no longer used and a programme of research into rehabilitation techniques. The three yearly Environmental Management and Monitoring Plan developed by WMC Ltd, and approved by the State Government, specifically addresses rehabilitation practices and issues as one of the integral elements of the plan. This plan is publicly available.

General issues

Senate inquiry

A Senate inquiry into uranium mining and milling has added to the pool of information on uranium mining and milling. The terms of reference of the Senate inquiry are:

- The environmental impact of uranium mining and milling in Australia and the effectiveness of environmental protection and monitoring in relation to existing and previous Australian uranium mining operations.

- The role of the Supervising Scientist in monitoring Australian uranium mining and milling activities.

- The health and safety implications of uranium mining and milling for workers at mining and milling sites and mining operations.

- The health and safety and other effects of uranium mining and milling on communities adjacent to mining and milling sites and communities on existing or planned transport routes for uranium ore and uranium wastes.

- The effectiveness of Australia's bilateral agreements with countries importing Australian uranium in ensuring that Australian-sourced uranium is not used in military nuclear technology or nuclear weapons testing activities.

- The volume and location of Australian obligated plutonium currently in existence in the international fuel cycle (produced as a result of the use of Australian uranium), in what form it exists (for example, separated or in spent nuclear fuel) and its intended end use.

Two reports have been released from this inquiry. The committee tabled its report in May 1997. The majority report (Coalition/Labour) concluded "that the principal findings of the Ranger Uranium Environmental Inquiry (the Fox Report) has [*sic*] been vindicated by two decades of experience". Fox stated that: "the hazards of mining and milling uranium, if those activities are properly regulated and controlled, are not such as to justify a decision not to develop Australian uranium mines". A minority report from Senators opposed to uranium mining stated that the uranium industry should be closed down, or failing that, 33 recommendations "aimed at ameliorating the harm done to Australia by continuing this deadly trade" should be implemented. A Government response to the Inquiry is in preparation.

Environmental issues – perception is reality

Lobby groups which are opposed to uranium mining on philosophical and/or environmental grounds, are adept at cultivating negative impressions of the industry. Their strategy is to propagate mistrust of the industry and government, and instil fears of environmental vandalism within the Australian public. The uranium mining industry responds by publicising the details of their environmental management programmes, and in recent years, some companies have opened their operations to public scrutiny. Lifting the veil of secrecy has forced the lobby groups to be accountable for their claims. Similarly, the lobby groups have forced the industry and government to be cognisant of public opinion. The lobby groups play an important role in pressuring the industry and government to continually improve environmental performance.

Often, issues are raised which may seem trivial or of no consequence to the industry or government. The perception of these issues by the Australian public will be flavoured by lobby groups, and/or individual social and moral values. Hence, the environmental risks associated with the issue may be perceived by the public to be far more severe than those calculated by environmental scientists. The natural reaction of the industry and government has been to either accuse the public of ignorance and ignore the issue, or attempt to explain to the public why the risks are small and ignore the issue. Both of these strategies simply serve to fuel the public "outrage". To the public, who are not experts in environmental or radiological science, perception is reality. The Australian uranium mining industry and government are slowly learning that the management of public outrage is as important as the management of environmental risks. Through consultative mechanisms, especially in the Alligator Rivers Region of the Northern Territory, issues of perceived and real environmental significance are openly discussed, and mitigating measures are designed and implemented. The management of public outrage has always been an issue in Australia. Only recently has the industry and government realised how to manage it effectively, although much remains to be learned.

Thus the environmental issues associated with uranium mining in Australia are driven by both public perception and technical consideration. In many cases, matters of perceived significance to the public coincide with those identified by industry and government as requiring attention. One which fits this description, as alluded to in the previous section, is the efficacy and equity of the current regime of environmental regulation controlling uranium mining. Other issues of note are: management and disposal of mine and processing wastes (tailings); and water management.

The Ranger Mine is located in a monsoonal environment receiving approximately 1 400 mm of rain during each wet season. Excess water collected during the wet season is managed in a sensitive World Heritage listed area.

The Olympic Dam mine is located in an arid environment and draws water from the Great Artesian Basin. The use of artesian water by the mine has been an important part of the EIA process and is subject to regulation and monitoring to ensure that no unacceptable impacts, for example on associated mound springs, occur.

The substantive rehabilitation of the Nabarlek uranium mine was completed in 1995. Environmental monitoring of the site continues, however the time frame for cessation of monitoring and release of the site to its traditional owners is yet to be determined. Other uranium mines developed in the 1950s and 1960s were rehabilitated or subject to hazard reduction works in the 1980s, and need continuing monitoring and maintenance.

Environmental research priorities

The Environmental Research Institute of the Supervising Scientist (ERISS) has a history of over twenty years of environmental research pertaining to the Ranger uranium mine. ERISS has developed local standards and environmental and biological monitoring protocols, and highlighted the need for internationally agreed generic Codes of Practice or Guidelines on environmental aspects of uranium mining. Some areas of research which should be given priority in order to facilitate the development of such codes of practice are:

- Close-out criteria for mines.

- Standards against which rehabilitated mines may be assessed to determine the acceptability or otherwise of rehabilitation measures including the types and period of monitoring.

- Environmental integrity of techniques for the final disposal of tailings. Comparison of below and above ground methods and potential new technologies such as pre-disposal tailings treatment.

- Development of drainage systems to improve the integrity of tails containment, particularly in regard to solutes which are potential contaminants of aquifers and wetlands.

- Water resource protection.

- Development of protocols for sensitive environmental monitoring systems in aquatic and terrestrial environments, which are non-destructive and give reproducible results. Protocols should embrace biological ecosystem monitoring rather than be restricted to the measurement of chemical, physical and radiological parameters.

- Atmospheric releases (radon, dust, atmospheric mill effluent).

- Assessment of effects on flora and fauna of atmospheric contaminants, methods for reducing atmospheric releases, development of monitoring protocols including the identification of critical species and ecosystems.

- Biological water treatment processes.

- Investigation of the capabilities and limitations of wetland filters to cleanse contaminated water. Development of design and optimisation principles for wetland filters. Assessment of rehabilitation requirements for wetland filters.

REFERENCES

DPIE (1996), *Submissions to Senate Select Committee on Uranium Mining and Milling 1996* by the Department of Primary Industries and Energy (unpublished).

WMC Ltd (1996), *Submissions to Senate Select Committee on Uranium Mining and Milling 1996* by WMC Ltd (unpublished).

• Brazil •

URANIUM INDUSTRY AND ENVIRONMENTAL ASPECTS

In Brazil, experience regarding environmental aspects has been taken from the first and unique uranium mining and milling facility which was first put into operation in 1982. The Uranium Mining and Milling Facility of Poços de Caldas, located at one of the most important alkaline intrusions in the world, is the only uranium mine operating in Brazil. The open pit covers an area of about 2.0 km^2. Host rocks consist of a series of hydrothermally and metasomatically altered intrusive bodies and flow of volcanic to sub-volcanic phonolites, varying in the proportion of mafic minerals, and nepheline syenites (Waber *et al* 1991).

It has been estimated that 94.5×10^6 tonnes of rocks were removed due to mining operation (Fernandes *et al.* 1993). Only 2% of this amount was processed to recover uranium. The rest is contained in the waste rock piles. The ore is processed using a sulphuric acid leaching plant. The overall U_3O_8 production, after 10 years of operation, has been estimated to be in the order of 1 000 tonnes.

Drainage from the mine is retained inside the open pit by an engineered trench. The drainage from the waste dumps is collected in artificial ponds and pumped out to the mine and mixed with the mine drainage. Those waters are pumped to the neutralisation unit for treatment. This produces a sludge that is deposited in the tailings impoundment along with the mill tailings. The effluent of the tailings dam is treated with $BaCl_2$ to precipitate the soluble $^{226 \& 228}Ra$ as $Ba(Ra)SO_4$.

Two hydrographic basins cross the mine site. Waters from the two rivers are intensively used for crop irrigation and watering cattle. Recreational fishing also takes place in these rivers. The annual average temperature is 19°C, ranging between 1°C and 36°C. The total maximum precipitation is 1 700 mm, with more than 120 days of rain per year (Amaral *et al.*, 1988).

The impact of the untreated acid rock drainage release to the environment was first assessed by Amaral *et al.* (1988) . The annual dose due to ^{226}Ra and ^{238}U from the waste rock leaching amounted to 0.59 mSv. The radiological impacts of the mining and milling industry operation as a whole has been assessed by Amaral (1992). The author has estimated an annual effective dose equivalent in the range of 3.7×10^{-2} to 1.4×10^{-1} mSv for hypothetical critical groups potentially exposed through eating fish, irrigated vegetables, milk, beef and water ingestion, as well as water immersion and exposure to contaminated sediments. The dose was assessed considering all long-lived radionuclides from the uranium series released with the liquid effluents of the chemical processing, mine drainage and acid rock drainage.

The impacts of Rn exalation and dust emissions from the facility could not be distinguished from the natural dose received by the local population. No increment on aerosol concentration was observed and regarding radon, no gradient was found. The natural exposure to Radon for local inhabitants is estimated to be in the range of $3.7 - 11.4$ mSv.y^{-1} (Amaral *et al.*, 1992).

Veiga *et al.* (1997) used a screening approach to study the comparative impacts of non-radioactive and radioactive pollutant emissions to the environment via liquid effluents. They report that all contaminants identified as significant for further investigations were related to non-radiological health effects. They are: Mn, F and U. Despite the fact that uranium is a radioactive element, its chemical toxicity to humans was of greater concern than its radiological effects.

A comprehensive research project is being carried out to investigate the appropriate remedial actions to be adopted during the mine facility close-out. Three main pollution sources are being evaluated: waste rock dumps, the tailings dam and the open pit. For all of these entities, acid drainage, resulting from the presence of sulphidic material in the host rock, is the driving force for the mobilisation of both radioactive and non-radioactive pollutants .

Fernandes *et al.* (1996a) have reported that in a worst-case scenario comprising the termination of the chemical treatment of the liquid effluent released from the tailings dams (with no remediation such as a cover over the tailings taken into account), the estimated doses to be incurred by the critical group would vary in the range of 8.1-8.5 mSv.y^{-1} (conservative approach) and from 0.48-0.62 mSv.y^{-1} (non-conservative approach). The radioactive species ^{210}Pb and ^{210}Po would account for more than 80% of the total dose.

Fernandes *et al* (1996b) have studied alternatives to the acid drainage remediation. It has been estimated that more than 1 000 years would be needed for the pyritic material to be naturally consumed resulting in termination of the acid drainage generation. The radioactive species ^{238}U, ^{234}U and ^{230}Th would account for 97% of the total dose (estimated to be about 0.83 mSv.y^{-1}).

These results show the potential importance of the liquid effluents from uranium mining and milling to the individual dose, particularly where pyrite is present in the ore and the local meteorological conditions promote development of acid mine waters. Therefore, for planning environmental management, site specific, as well as climatological and geochemical conditions must be taken into account. The socio-economical aspects of the site may also be very important.

The experience obtained from these studies leads to the implementation of a methodology based on a mass balance of the radionuclides activity, together with a risk assessment at the non-radioactive mining/milling facilities. At this site uranium or thorium occurs in the ore from which other commodities are the main product. A new project is being developed in Brazil with the aim of giving priority to environmental management procedures at the installations. For those installations one alternative is the recovery of uranium from the waste, even though it would only be recovered as a by-product.

Treatment of liquid effluent from uranium mining and milling operation at CIPC in the Poços de Caldas plateau – Minas Gerais

The Complexo Minero Industrial do Planalta de Poços de Caldas, CIPC, is one of the industrial plants of Industrias Nucleares do Brasil S.A., INB, responsible for nuclear fuel cycle activities in Brazil.

CIPC is the only plant operating in the country for the production of uranium concentrate as ammonium diuranate $(NH_4)_2U_2O_7$ with about 75% of uranium content. It is located in an area of approximately 15 km^2 in the Poços de Caldas Plateau, state of Minas Gerais, within the watershed boundaries of the most important hydrographic basins of the region, the Antas and Verde rivers. The mine considered in this study is situated in the tropical climatic zone characterised by a large average precipitation of 1 800 mm, mainly occurring during the summer period extending from October to March. Annual average temperatures range between 7.5°C minimum and 25.9°C maximum.

The mining and milling operations include crushing, grinding, chemical processing, and yellow cake packaging. The mine pit of 1.2 km in diameter was mined out from maximum elevation above sea level from 1 480 m to 1 304 m, presenting a height of 176 m. So far, 2.0 x 10^6 metric tons of ore have been exploited resulting in 43.0 x 10^6 m^3 of waste material deposited as rock piles over an area of 1.73 km^2. Since 1993, the waste rocks have been returned to the mine pit. From the beginning of operation in 1982, action has been taken to minimise the environmental impact of these piles. These activities were later strengthened by management programmes designed for monitoring water draining from the piles.

The flow sheet for CIPC's chemical processing includes conventional sulphuric acid leaching of ore, separation of solid phase by filtration, liquor clarification, organic solvent extraction, aqueous sodium chloride re-extraction and alkaline precipitation of ammonium diuranate. This plant, with a nominal processing capacity of 2 500 metric tons of ore per day, generates solid waste similar to ground ore and liquid effluent from which the uranium has been extracted. The solid waste, recovered as a pulp, is alkalised with lime to pH 9.0 and directed to the tailings pond. The liquid effluent is neutralised to pH 4.5 with limestone and thereafter to pH 9-10 with lime and sent to the tailings pond.

The main sources of liquid effluent at the CIPC are the mine pit, waste rock piles, ground ore stockpile yard and mill tailings pond.

Mine pit

Uranium occurrences in the plateau were first observed in 1948, with geological investigation and prospection initiated in 1964. Considering the economical exploration of the ore, activities of research and prospection have been developed in the mine area since 1971, with 14 000 m of drilling and 2 500 m of tunnels, leading to the choice of open pit mine operation. Mining started in 1977 with mine scouring, and the first ore pile was constructed in 1981. The mine was essentially formed by three mineralised bodies, named A, B and E in relation to their individual geological characteristics. Waste rock includes mine material with a soluble U_3O_8 content of less than 200 ppm, the mill cut off grade. According to the source, the waste rock is disposed of in different areas designed for this purpose around the mine pit. Mine material was removed by breaking individual blocks of ore 5 m x 5 m x 2 m into the maximum rock size of 1 m. The ore was then sent to primary crushing and the waste rocks disposed in piles. The deposition was carried out using either progressive end-dumping or horizontal layers, depending on the mining dynamics.

Acidic mine drainage, which originates from several points inside the pit, is first directed to a trench of 80 000 m^3 capacity. It is then pumped to ponds A1, B1 and B2, thereafter to the lime treatment station. The chemical composition of the acid waters is similar to that encountered for acidic drainage from waste rock piles which is discussed in the next section.

Waste rock piles

The first choice for waste rock disposal was the construction of a single pile around the tailings pond. This choice was not made because of the turfy terrain found at the site. The search for an adequate place focused primarily on the stability of the substrate, and then on the economical aspects of haul distance and topography.

Overview of the characteristics of waste rock piles (WRP)

WRP	Volume $10^6 m^3$	Surface area $10^4 m^2$	Main source of material
1	4.4	25.5	Overburden
3	9.8	20.5	Overburden
4	12.4	56.9	Waste material from body B+overburden
7	2.4	5.3	Overburden
8	15.0	64.4	Waste material from body (B+E)+overburden
WRP inside the mine pit	0.56	9.87	Waste material from body E

Among existing waste rock piles, the most significant in terms of interaction with the environment are piles 4 and 8. This is due to their location, origin and quantity of deposited material. Environmental hazards and associated management actions for physical, chemical and biological stabilisation are similar for both piles. Therefore studies for their decommissioning now concerns only pile 4.

Pile 4 was constructed over the Consulta valley in an area near the mine pit, since field examination of the geological and geotechnical features did not define any instability or the possibility of embankment rupture. No turf or soft soil was found at the banks of the creek and tributaries. The valley is formed exclusively by a continuous surface of saprolite, an altered soil derived from alkaline rocks of high hardness and resistance. Over this surface lies a 0.3 m thick layer of limonite gravels, clay and sand. Maximum discharges of 31.5 and 11.1 l/sec were measured for the Consulta and its tributaries, respectively. Pile 4 contains $12.4 \times 10^5 \text{ m}^3$ of waste rock and mine overburden distributed over an area of 0.57 km^2, with a maximum embankment height of 90 m. The waste rock came mostly from mining of the body B, which was primarily constituted of a mass of breccia in a pipe configuration. It formed from the syenitic intrusion commonly seen in the mine area. The matrix exhibited a tingualite texture impregnated by hydrothermal materials like pyrite, fluorite, uranium minerals, molybdenum and zirconium, as addition to galena, sphalerite and baryte in small amounts. Kaolin has been noted filling the fractures or even disseminating into the porous zone of the breccia. The uranium present in the ore matrix was either from primary origins linked to ascendant hydrothermal process that permeated the breccia bodies or from secondary reconcentration due to the action of a reoxidation process. Dimensions of body B were 400 m wide x 500 m long and 400 m deep. It has been mined out to a height of only 1 332 m above sea level for economical reasons. The uranium and molybdenum concentrations were 800 and 1 600 $\mu g \ g^{-1}$, respectively.

Pile 8 was constructed over the Cercado valley following the same selection criteria described above. This pile contains $15 \times 10^6 \text{ m}^3$ of overburden material and waste rock from ore bodies A and E.

Body A presented a similar matrix to that encountered in body D breccias. It was constituted by tinguaite/phonolite mass where potassic feldspar and sericite were the dominant minerals. Pyrite, fluorite, uranium minerals, molybdenum and zirconium also commonly occur. Kaolin, a typical product of alteration, was frequently found in fractures or in disseminations throughout the breccia. Approximated dimensions of the body were 300 m wide, 400 m long and 250 m deep. It was mined to a height of 1 320 m above sea level. Uranium and molybdenum contents were 700 and 1 500 $\mu g \ g^{-1}$, respectively.

Body E is disposed around breccia bodies and corresponds to the rocks adjacent to the breccia where secondary uranium concentration occurs. The dominant lithology is very highly fractured tinguaites/phonolites, which have also been strongly altered by hydrothermal activity. The secondary factors responsible for uranium concentration in this area are the increased tectonic and thermal effects.

These features promoted the weathering action and consequent migration of uranium impregnating solutions, by advance "per decensum" redox fronts acting throughout the orebody. Under such conditions, uranium is leached and redeposited in reduced zones. It occurs as films disposed along fractures or disseminated in some zones of high porosity. Nodular concentrations were frequently found where fractures cross and in areas where dekaolinization of rock was more intense. In this case, because of the differential geochemical migration cycle, the molybdenum and zirconium contents are usually low, particularly as compared to orebodies A and B. The average uranium content was 1 200 $\mu g \ g^{-1}$.

Before filling the valleys with waste rock, drainage of the bottom surface was prepared by constructing deep drains with waste rocks covered by transition material and clay. Taking into account the physical stabilisation of the deposits and the reduced alteration in Consulta and Cercado creeks, a deviation of about 500 and 1 500 m of respective courses lying inside the area of the waste rock deposition had been performed to a downstream point. The surface of the relocated material is covered with a 30 cm layer of compacted clay to prevent rain water percolation through the waste rock. Otherwise rainwater could enhance the volume of contaminated drainage waters. The final slope of the pile ranged from 0.5 to 1%. The rain waters falling upon this platform is conveyed to the environment through drainage channels. To stabilise the deposited material, alternatives for reforestation have been tested for protection against wind and rain erosion and, to a certain degree, the penetration of humidity into the piles.

Monitoring of the water in Consulta and Cercado valleys showed a significant increase in the uranium and other dissolved solids due to the influence of dissolved metals in the waters draining from the waste rock piles. These waters are usually acidified by the oxidation of pyrite found in the waste rocks, and have the following radiological and chemical characteristics: in Bq/l, ^{226}Ra=0.30 ^{228}Ra=0.20, ^{238}U=79.3; in mg/l, Mn=80, Al=170, Fe=2.1, Ca=95, SO4=1300, F=100; and a pH=3.0-3.5. Permissible limits for releasing water into the environment at the monitoring site are: in Bq/l ^{226}Ra=1.0, ^{238}U=1.0; in mg/l, Mn=1.0, Al=0.1, F=10. Therefore a drastic reduction of dissolved U, Mn, Al and F is necessary. This has been accomplished by collecting such waters in ponds, then pumping them to the lime treatment station. The precipitated solids are directed to the chemical processing plant or to the mill tailings pond. The overflow liquid is sent to the solid settling ponds before discharge to the environment.

The results of an ongoing chemical and radiological characterisation of waste rock pile 4 is given in the following table.

Average composition of waste rock pile 4 material

Element	Concentration ($\mu g.g^{-1}$)	CV(%)	Element	Concentration ($\mu g.g^{-1}$)	CV(%)	Element	Concentration ($\mu g.g^{-1}$)	CV(%)
As	59.9	42	K	97 380	19	Sm	26.8	63
Ce	1 044	79	La	803	88	Ta	7.44	27
Co	1.54	124	Lu	1.54	45	Tb	3.21	46
Cs	0.68	47	Na	1 057	39	Th	90.3	36
Eu	7.46	58	Nd	266	74	U	219	52
Ra	30 534	58	Rb	270	16	Yb	13.1	45
Hf	38.3	76	Sc	1.54	54	Zn	266	76

Ore stockpile yard

The dimensions of the ore stockpile yard are 175 m x 415 m and its pile capacity is 200 000 metric tons of ground ore. The rainfall in the yard is collected in ponds B1/B2 and pumped to the lime treatment station.

Mill tailings pond and barium chloride treatment system

The uranium ore, after having been extracted from the mine, is crushed in three stages. At the last stage, pyrolusite and phosphate rock are added. The first assures an oxidising medium and the second precipitates the zirconium solubilized during acid leaching. Crushed materials are ground with the addition of water to a particle size 85% below 48 mesh Tyler (1.19 mm). The ore pulp is sent to a sulphuric acid leaching process with agitation and heating. Uranium bearing liquor is separated from solid by filtration. After clarification by settling of the suspended solids and activated carbon filtration, the liquor is sent to the solvent extraction area. Uranium is re-extracted by adding a sodium chloride solution and precipitated by adding ammonium hydroxide. The amonium diuranate precipitate is filtered. The pulp is sent to a spray dryer and then packaged. Solid waste from acid leaching is collected as pulp, and neutralised with lime to pH 9. It is then sent to the mill tailings pond where liquid solid separation occurs by gravity settling.

Uranium free acidic liquor is treated with limestone to pH 4-4.5, with lime to pH 9-10, and then sent to the mill tailings pond. During elevation of the pH, a considerable amount of iron, aluminium and manganese hydroxide, and calcium phosphate, fluoride and sulphate are formed and deposited in the pond.

The tailings pond is contained by a dam 10 m wide, 435 m long at the top and with a maximum height of 42 m. The average mill effluent discharge rate is $0.15 m^3/s$ and maximum flush rate is $0.45 m^3 s^{-1}$. The capacity of this pond is $2.39 \times 10^6 m^3$, with a surface area of $0.23 km^2$. It is located in the Verde River basin, next to the Antas River basin. These two basins form 80% of the hydrologic system of the Poços de Caldas Plateau making them important to the social and economic well-being of the region. As a result it was imperative to carry out a careful study to select an appropriate disposal site. The Soberbo creek drainage area was selected based on the following favourable considerations: topography which avoided construction of massive dam; a small drainage area to the dam; no significant ground water drainage; a favourable subsoil profile in relation to either supporting or retention capacity; possibility of capturing drainage water through the dam and foundation and directing it into a well defined and restricted area; located near the present tailing site thereby reducing transport and deposition costs. The pond area is underlain by more than 20 m of red clays with sparse emergence of the rocky substrate. The substrate is composed essentially of pseudo leucite-tinguaites, intercalated with microfoites. The presence of compacted clays was a strong factor in choosing the area. They form a thick natural barrier preventing the loss of liquids impoundment either to the Antas basin or to the rocky substrate. Because of the low permeability ($10^{-7} cm s^{-1}$) of the thick clay layer, no additional clay or artificial liner was necessary.

At a site below the dam there is a water treatment station where barium chloride may be added to insolubilise radium in the overflow stream from the pond. After the barium radium sulphate crystals settle in two ponds, the overflow is released into environment. A total of 2 452 580 tons of solid waste were deposited in the mill pond.

Chemical and radiological characteristics of mill solid waste

Element	Concentration	Element	Concentration
ZrO_2	0.15 %	Fe_2O_3	3.9 %
MoO_3	0.03 %	Mn	0.02 %
Al_2O_3	23.4 %	P_2O_5	0.09 %
K	11.2 %	U	0.018 %
SiO_2	54.0 %	Th	0.004 %
CaO	0.25 %	^{226}Ra	2.5 Bq g^{-1}
SO_4	2.3 %	^{226}Ra	1.4 Bq g^{-1}
S	0.5 %	^{210}Pb	3.4 Bq g^{-1}

CIPC's environment management

The environmental management of CIPC aims at evaluating and minimising the impact of the installation activities. For this purpose, the management plan has established an environmental and effluents monitoring network, a tailings treatment and retention system, a pumping network and a lime treatment station for acidic drainage waters, and a well-established procedure for controlling the pollutant release to acceptable levels. The overall management plan comprises studies for stabilisation of pollutant sources, rehabilitating sites to restore the environment or make them suitable for other uses.

Efficiency of effluent treatment

An annual average volume of about $2.2 \times 10^6 \, m^3$ of water have been collected from waste rock piles 4 and 8, the mine pit and ore stockpile yard, and then pumped to the lime treatment station.

In 1995, CIPC produced 121 tons of U_3O_8 (yellow cake), processing 135 400 metric tons of ore. This operation generated almost the same amounts of solid tailings and 248 730 m^3 of uranium depleted liquor (raffinate) which has been treated with limestone/lime and sent to the tailings pond. A total of about $1.48 \times 10^6 \, m^3$ of overflow from pond, containing no significant suspended solid, has been released into the environment after removing radium using the barium chloride treatment system.

Chemical and radiological characteristics of milling liquid effluent

Physico-chemical parameters	Raffinate	Mill tailings pond overflow	Environment release	Release limit
Concentration		**(mg/l)**		
Mn	1 620	<0.05	8.8	1.0
Fe	1 450	–	–	–
FE	730	–	–	–
Fe	–	–	0.2	15.0
Al	548	–	9.8	0.1
Ba	–	–	0.3	–
MoO_3	13	–	–	–
ZrO_2	<5	–	–	–
Ca	978	596	406	–
MgO	80	–	–	–
SiO_2	25	–	–	–
SO_4	38 500	1 400	984	250
F	200-300	5-10	5.6	10
P_2O_5	790	–	–	–
Free acidity (as H_2SO_4)	20 000	–	–	–
PH	1-1.5	9-10	7.3	–
Radionuclides		**Activity (Bq l^{-1})**		
U	6.3	0.4	0.32	0.4
^{228}Ra	15	3	0.2	0.2
^{210}Pb	65	<0.2	<0.2	0.2
ThO_2	13	<0.02	<0.02	0.2
^{226}Ra	<1.7	<0.2	<0.2	0.2

The limits were established by Comisseo Nacional de Energia Nuclear, CNN, CONOMA Resolution No. 20 Art. 2 of 18/06/86 and CONOMA Resolution No. 20 Art. 4 of 18/06/86.

Estimated investment and maintenance costs of the acid drainage treatment system

The main investments and maintenance costs of collecting, pumping and treating acid mine drainage were estimated, taking into account CIPC's decommissioning.

Items	Costs (USD)
1. Constructions for diverting creeks out of waste rock piles, piles surface covering with clay and regularisation, surface drainage net construction	645 045
2. Construction of acid drainage collecting and pumping system to lime treatment area	540 000
3. Equipment operation and maintenance of pumping, treatment system and water drainage monitoring (3 700 man x hours/month)	32 000
4. Chemicals/year	
Lime	277 600
5. Laboratory material	22 000
6. Electrical energy/year	22 000
Total cost	
Investment (1 + 2)	*1 185 045*
Operation/year	*705 800*

Studies envisaging a definitive stabilisation of drainage sources

Despite the fact that the approach adopted by the present management to minimise the environmental impact is considered to be highly efficient, a definitive solution is still being sought for the elimination of the potential pollution source. Most of the actions to identify a final solution encompass identification of the physical, chemical and biological aspects of the problem, as well as its temporal evolution.

A complete survey of the former attributes of the areas presently occupied by the deposits, including studies of the fauna, flora, as well as surface and underground hydrology, is necessary for their rational reintegration into the environment at the time of mine decommissioning.

REFERENCES

Amaral, E.C.S. (1992), *Modificaçao da Exposiçao à Radiaçao Natural Devido a Atividades Agricolas e Industriais numa Area de Radioatividade Natural Elevada no Brasil*, Ph.D. Thesis – Instituto de Biofisica Carlos Chagas Filho, Universidade Federal do Rio de Janeiro, Brazil, 130 pp.

Amaral, E.C.S., Godoy, J.M., Rochedo, E.R.R., Vasconcellos, L.M.H. & Pires do Rio, M.A. (1988), *The Environmental Impact of the Uranium Industry: Is the Waste Rock a Significant Contributor?*, Radiation Protection Dosimetry, Vol. 22, No. 3, pp 165-171.

Amaral, E.C.S., Rochedo, E.R.R., Paretzke, H.G. & Franca, E.P. (1992), *The Radiological Impact of Agricultural Activities in an Area of High Natural Radioactivity*. Radiation Protection Dosimetry, Vol. 45., No. 1/4, pp 289-292.

Fernandes, H.M., Prado, V.C.S, Veiga, L.H., Freitas, P., Amaral, E.C.S. & Bidone, E.D. (1993), *Risk Management in Environmental Pollution: A Case Study on the Uranium Mining and Milling Facilities of Poços de Caldas, Minas Gerais, Brazil*, Proc. of the Latin American Section of the American Nuclear Society Symposium, Rio de Janeiro, Brazil.

Fernandes, H.M., Franklin, M.R, Veiga, L.H., Freitas, P. & Gomiero, L.A. (1996a), *Management of Uranium Mill Tailings: Geochemical Processes and Radiological Risk Assessment*. J. Environmental Radioactivity, Vol. 30, No. 1, pp 69-95.

Fernandes, H.M., Veiga, L.H., & Franklin, M. (1996b), *Application of Environmental Management Concepts to a Nuclear Installation: A Study Case on the Uranium Mining and Milling Facilities of Poços de Caldas – MG – Brazil*, Proceedings of General Congress of Nuclear Energy (CGEN), Rio de Janeiro, Brazil.

Veiga, L.H., Amaral, E.C., Fernandes, H.M. (1997). *Human Health Risk Screening of Radioactive and Non Radioactive Contaminants Due to an Uranium Industry*. J. Environmental Radioactivity (in press).

Waber, N., Schorscher, H.D. & Peters, T. (1991), *Mineralogy, Petrology and Geochemistry of the Poços de Caldas Analogue Sites, Minas Gerais, Brazil*. Technical Report TR 90-11. Swedish Nuclear Fuel and Waste Management Co., 513 pp.

• Bulgaria •

HISTORY OF PRODUCTION CLOSURE

In 1992 the Government of the Republic of Bulgaria passed Ordinance Number 163 which decreed an end to uranium production activities and authorised the liquidation of uranium production and processing sites. An Inter-institutional Experts Council with representatives of all organisations formerly involved in uranium production was formed. The task of the Expert Council is to evaluate all identified sites and prepare a list of those sites that are eligible for environmental restoration under the programme. The Council is also responsible for co-ordinating and directing all activities for the liquidation of uranium production.

An additional Government Act is planned to define the structure of the organisation. The objective of this act is to improve the organisation, as well as provide a plan for the financial resources required for implementing the remaining reclamation work which includes technical and biological reclamation, decontamination, management and monitoring of waters.

Work conducted under this programme involved defining the environmental parameters of areas affected by uranium production. A total of 128 studies including radio-ecological and hydrological surveys were conducted by expert teams. The studies were conducted to define environmental problems and evaluate the feasibility of alternative proposals for liquidation. The stages of liquidation and restoration activities for all sites were divided into 5 activities: 1) technical liquidation; 2) technical reclamation; 3) biological reclamation; 4) decontamination of mine and flowing surface waters and 5) monitoring.

By April 1995, technical liquidation had been approved for 61 uranium production sites. The plans called for the implementation of this work by September 1995 (with the exception of the 2 processing sites). By the end of March 1995, 70% of the work had been implemented at a cost of 50 million Deutsche Marks (DM). This sum is considerable, given the economic situation of the country.

Of the sites remaining to be restored, the environment problems are greatest in Bukhovo, Eleshnitza, Plovdiv District, Haskovo District and Smolyan. Problems at these sites are associated with tailings ponds, contaminated ground water, mine waste dumps and lack of adequate monitoring systems.

ENVIRONMENTAL ASPECTS

The current activities involve technical and biological reclamation (i.e. revegetation). Environmental monitoring is continuing during reclamation activities.

The suspension of uranium production and the reclamation of the areas related to uranium mining activities resulted in the following tasks: selection of a cost effective technology for treatment of mine and surface waters contaminated with radionuclides; selection of equipment for environmental monitoring of uranium production sites undergoing closure; design and/or selection of cost effective methods for treating and restoration of tailings ponds and waste dumps associated with uranium production.

• Canada •

THE IMPACT OF CANADA'S ENVIRONMENTAL REVIEW PROCESS ON NEW URANIUM MINE DEVELOPMENTS

Canada introduced an environmental assessment process in the mid-1970s. It was designed to ensure that the environmental consequences of all project proposals with federal government involvement were assessed for potential adverse effects early in the planning stage. In 1984, a Guidelines Order was approved to clarify the rules, responsibilities and procedures of the Environmental Assessment and Review Process (EARP) that had evolved informally under earlier Cabinet directives.

The EARP process remained largely uncontested until 1989/90, when two decisions by the Federal Court of Appeal effectively converted the Guidelines Order into a legal requirement for rigorous application. The Supreme Court of Canada upheld the constitutionality of the EARP Guidelines Order in 1992, rendering compliance with the Order by all federal government decision-makers a mandatory requirement.

Canada became the world's leading producer and exporter of uranium during the late 1980s. Since then, the Canadian public has become sensitised to numerous issues concerning environmental degradation, from the Chernobyl accident to ozone depletion. In 1991, during this period of increasing awareness, the Atomic Energy Control Board, the federal nuclear regulator, referred six new Saskatchewan uranium mining projects for environmental review, pursuant to the EARP Guidelines Order.

Both federal and provincial governments and the uranium industry expressed concern about the potential impacts of an apparently burdensome public environmental review process on the development of new uranium mining projects. However, the successful advancement of three such proposals through the process has alleviated much of the uncertainty and has confirmed that Canada's uranium-producing operations can meet high environmental, health and safety standards.

The public review process provided an extremely valuable focus on aspects of these developments that needed to be addressed by proponents and regulators. It has helped to demonstrate that new uranium mining projects are being developed in a responsible manner, after full consideration has been given to the potential impacts and public concerns associated with these facilities. The lessons learned in Canada could well prove to be useful in other jurisdictions where considerations are being given to the development of new uranium production projects.

Introduction

The establishment of the Canadian government's process for assessing the environmental impact of major new projects was precipitated in part by a growing appreciation among Canadians that a balance must be achieved between the activities that promote economic growth and the preservation of the natural environment. Governments have become increasingly aware that economic growth must be managed in a manner that is sustainable in the longer term. When public confidence in an environmental assessment (EA) process is achieved, and economic growth is seen to be managed responsibly, the continuation of development is supportable by everyone. When a viable process is accepted in one jurisdiction, the approach may quickly become the norm in others.

This report provides a brief examination of the evolution of Canada's EA process, focusing primarily on its application to the uranium mining industry. The new environmental assessment legislation is compared with the previous federal EA regulations under which all the current uranium mining projects are, or will be, reviewed. Some recent examples are provided of the nature and scope of the reviews undertaken related to uranium mining projects, and of their outcomes. A brief analysis of this recent experience, together with some observations about how the effectiveness of the process might be improved, may be useful for those considering adopting the Canadian model as part of the approval process in the development of their uranium resources.

Perceptions and public awareness

Canada has been a major producer of uranium since the inception of the industry in the 1940s. However, Canada's emergence as the world's leading producer and exporter of uranium in the mid-1980s coincided with a surge in anti-nuclear activity which sought to block further nuclear power growth. The prospect of uranium mining expansions in Saskatchewan, together with a Canadian public newly sensitised to environmental matters, afforded anti-nuclear advocates the perfect opportunity to discredit the uranium mining industry. In parallel, Canadians had become increasingly aware in the

1980s of the importance of maintaining a strong economy and a healthy environment. World events focused public concern on the need for policies and practices that promoted sustainable development.

Critics pointed to falling uranium prices, reactor cancellations, and lower uranium demand. The Three Mile Island and Chernobyl accidents together with the uranium mine-water spills in Saskatchewan became "proof" that nuclear power and uranium mining were environmentally dangerous. Detractors argued that new uranium production capability was not needed. A public letter-writing campaign was launched to persuade governments to reject new uranium mining proposals on environmental grounds. When it became clear that industry intended to proceed with new mining proposals, the anti-uranium lobby called for a review of uranium mining in northern Saskatchewan. When this failed, critics turned to the EA process as a possible means of blocking new developments. They argued that only by imposing a new stringent review process could the environment and Canadians be protected. It was alleged that Canada did not have any critical environmental review mechanism in place for uranium exploration and mining activities. Clearly, this is not the case!

Regulatory framework for environmental review

Background

Canada is a federation of ten provinces. The respective jurisdiction of the federal government and the provinces is defined in the Canadian 1987 Constitution Act. Under the Constitution, the federal Parliament has the power "to make laws for the peace, order and good government of Canada", except for those areas which fall under exclusive provincial jurisdiction as specified in the Act. Many of these areas of federal jurisdiction are enumerated in the Act, such as defence, postal services, navigation, shipping, railways, international and interprovincial undertakings, money banking, and criminal law. Areas of provincial jurisdiction include such matters as natural resources, electricity generation, local works and undertakings, hospitals, education, property and civil rights, and the creation of courts and the administration of justice. There are also areas where the federal Parliament and provincial legislature share power.

The federal role in the nuclear industry

The federal government has overall jurisdiction over Canada's uranium industry by virtue of the Atomic Energy Control Act of 1946. All Canadian uranium operations are classified as nuclear facilities, and are controlled by the federal Atomic Energy Control Board (AECB) under the provisions of the Act and the regulations derived from it. Although uranium mining activities have always been subject to the requirements of the Act and its regulations, a strict licensing system has been in place since 1976. The AECB issues licences for the operation of uranium mines only after ensuring that they will not have a significant effect on the health and safety of the mine workers and the public, and after reviewing measures designed to ensure adequate protection of the environment. Furthermore, the AECB has always imposed on its uranium mining licensees the responsibility of reducing a miner's radiation exposure to levels as low as is reasonably achievable. Considerable advances in radiation protection regarding the risk of cancer have been made since the beginning of uranium mining in Canada. The AECB also regulates waste management practices to ensure that the radiological impact of Canada's uranium tailings does not pose any undue public health and environmental risks.

Provincial environmental protection

In Canada, all levels of government share responsibility for environmental protection. With the growing awareness that EA was becoming an essential tool for ensuring effective integration between economic and environmental imperatives, Ontario, already an important uranium producer, became the first province, in 1975, to establish its own Environmental Assessment Act. The following year, shortly before the province became Canada's leading uranium producer, the Government of Saskatchewan established an Environmental Impact Assessment Branch in its Environment Ministry, and in 1977 became involved in scrutinising and regulating its uranium mining industry in response to rising public concerns over proposals to develop new uranium mines in the Athabasca Basin. The provincial government appointed the Cluff Lake Board of Inquiry in 1977 and the Key Lake Board of Inquiry in 1979 to examine the Cluff Lake and Key Lake uranium mining projects, respectively, through a public hearing process. At these reviews, the public voiced concerns about environmental protection, the health and safety of workers, economic development, and the benefits to local communities. Both Boards of Inquiry found that the measures proposed by the uranium industry were adequate to protect environmental quality, safeguard occupational health and safety, and meet the requirements of Canadian and Saskatchewan law, regulations and policies in a satisfactory manner.

As a result of these and other non-uranium inquiries, the Government of Saskatchewan established a new Environmental Assessment Act in 1980, and created special units within its departments of the Environment and Labour to license and inspect uranium mines. Saskatchewan's EA review and regulatory processes have been structured to accommodate federal/provincial reviews for new uranium mining developments, as detailed later in this report.

The federal environmental assessment and review process

The federal EA process has paralleled the introduction of provincial procedures. Originally established by Cabinet Directive in 1973, and reaffirmed with some improvements in 1977, the federal Environmental Assessment and Review Process (EARP) was created to ensure that the environmental consequences of all project proposals with federal involvement were assessed for potential adverse effects early in the planning process. Specifically, EARP has been applied as an aid to predict the likely environmental effects of all proposals requiring federal involvement and decision-making. To oversee this process, the Federal Environmental Assessment Review Office (FEARO) was established, and operated within the federal Department of Environment.

In 1984, the Governor in Council approved a Guidelines Order to clarify the rules, responsibilities and procedures of EARP that had evolved informally under the earlier Cabinet Directives. The EARP Guidelines Order (EARPGO) set out a detailed assessment process for all proposals which would "include any initiative, undertaking or activity for which the Government of Canada has a decision-making responsibility". In this regard, "a decision-making responsibility" was interpreted as including proposals undertaken by any federal department, those likely to have an environmental effect on an area of federal responsibility, those for which the federal government makes a financial commitment, or those developed on lands or territories, including the offshore, administered by the federal government. While private-sector developments were not covered unless the development required federal involvement, for example the issuing of a licence, provincial regulations may apply to all projects within provincial boundaries. As discussed below, the current review of six uranium mining projects in Canada falls under the EARP Guidelines Order of 1984.

The EARPGO was written with a fair degree of flexibility and numerous areas were open to interpretation. Over time, it became clear that the EA process needed strengthening, clarification and reform; a revision of the process began in 1987. Nonetheless, the Order remained largely uncontested until 1989/90, when two decisions by the Federal Court of Appeal underscored the need for reform. In response to a challenge by environmentalists that approval of the construction of the Rafferty-Alameda and Oldman River dams in western Canada had been granted without an EA, the court ruled that what was thought to have been a non-enforceable guideline was, in fact, a legally enforceable law of general application that imposed additional duties to the existing responsibilities of federal decision-makers. This effectively converted the EARPGO into a legal requirement for rigorous application. Although these court decisions reflected a shift in public values, they also had a more significant impact in hastening reform. As the Order had not been drafted with strict legal interpretation in mind, the difficulty and uncertainty evident in its administration necessitated revisions in order to render the process effective, efficient, fair, timely and open. [See the section on The Canadian Environmental Assessment Act (CEAA) for a complete discussion of the new legislation].

Independent public environmental assessment and review panels

Before addressing the issues surrounding EARP as they relate to uranium mine developments, it should be noted that all new uranium mining projects are automatically referred by the AECB for public review, and that the review of a uranium mining proposal requires the formation of an EARP panel. As will be discussed later, much of the concern expressed by government and industry relates to the activities of these independent public review panels.

In April 1991, six uranium mining projects were referred by the AECB for public review. A five-member *Joint Federal/Provincial Panel on Proposed Uranium Mining Developments in Northern Saskatchewan* was appointed in August 1991 to assess the *Cluff Lake Dominique-Janine Extension (DJX)*, the *Midwest Joint Venture (MJV)* , and the *McClean Lake* projects. A four-member federal *Rabbit Lake Uranium Mine Environmental Assessment Panel* was appointed in November 1991 to conduct public hearings on the *Eagle Point/Collins Bay A & D Expansion*, also proposed in northern Saskatchewan. (See *Case Histories* for specifics on the recommendations of these two panels and the responses to the recommendations by government).

It should be noted that of the six uranium mining proposals referred in 1991 for panel review pursuant to the EARPGO, only three completed the review process before CEAA was proclaimed in January 1995. However, the remaining three uranium mining projects will be assessed under the EARPGO, since the review panels were appointed prior to CEAA being proclaimed.

Issues related to EARP, including overlap/duplication and scoping

As noted above, the Canadian and Saskatchewan governments both developed independent environmental assessment and review processes for those projects in which they would be involved. As well, each level of government also assumed specific regulatory responsibilities for various aspects of the development of uranium mining projects.

In accordance with the federal EARP Guidelines Order, the AECB established an automatic referral list for proposals such as new uranium mining facilities. The expansion or extension of existing facilities also became referable under various sections of the Guidelines Order, following an AECB screening process. When six uranium mining projects were proposed for development in Saskatchewan (see *Case Histories* below), each requiring an EA by a public review panel pursuant to EARP, it became clear that the existence of parallel processes at the federal and provincial levels would create unnecessary duplication.

In recognition of the similar legal requirements for a public input process, Saskatchewan and federal officials co-operated closely in establishing a joint federal/provincial process to review five of the six new uranium mining proposals. Saskatchewan agreed to include all but one of the proposals in a joint review process. As Cameco Corporation's *Eagle Point/Collins Bay* project had received provincial approval in 1988, there was agreement that it be excluded from review by the Joint Panel; it was reviewed only by a federal panel. This teamwork proved very beneficial and is highlighted by two examples of excellent federal/provincial co-operation.

First, establishing the Joint Panel's Terms of Reference (ToR) was crucial in limiting the overall scope of the review, so as to exclude issues such as nuclear non-proliferation and other national or international issues not directly related to the impacts of these uranium projects.

Second, the finalisation of guidelines for preparing Environmental Impact Statements (EIS) for the *Cigar Lake* and *McArthur River* projects was a major effort. The Joint Panel held scoping sessions to solicit views and opinions from the public and prepared guidelines with the help of a consultant. Circulated for comments in draft form in June 1992, these guidelines raised significant concern at both levels of government and in the uranium industry because they seemed to broaden the scope of the Joint Panel's review beyond that originally mandated. The guidelines also obliged proponents to submit marketing and operating information that could compromise their commercial positions, and sought data on social and health matters which were not readily, or in some cases legally, obtainable by the proponents.

The draft guidelines did not assign specific responsibility for the collection of data among the various expert groups, which might later be called upon to provide specialist testimony. They also suggested that governments provide guidance and information at the data-gathering stage and then criticise the results presented at public hearings, without realising that this might place governments in conflict of interest. Finally, they included questions about certain regulatory processes that were judged to be beyond the Panel's ToR. In summary, as drafted, the guidelines sought a great deal of information, significant amounts of which were not readily available and some of which was not directly relevant. All of these observations were submitted and the draft EIS guidelines were subsequently modified by the Panel.

General application and responsibilities under EARP

Before reviewing the progress of the first uranium projects through the panel process, it is useful to outline the role of the various players involved in an EARP review and how the process proceeds.

Application of EARP

As defined under EARP, any department, board or agency of the Government of Canada, including the AECB, may be considered an "initiating department", that is, a federal department or agency that has "decision-making authority" for a project. The EARPGO requires all initiating departments to consider the environmental impacts of proposals as early as possible.

Section 12 of the EARPGO obliges initiating departments to assess all projects, including those proposals with potentially significant adverse environmental effects. When it is determined that impacts may be significant or unknown, proposals are referred to the Minister of the Environment for independent panel review. However, Section 13 of the EARPGO states that "Notwithstanding the determination concerning a proposal made pursuant to Section 12, if public concern about the proposal is such that a public review is desirable, the initiating department shall refer the proposal to the Minister for review by a Panel".

Upon receipt of a panel report with recommendations, it is the responsibility of the initiating department to decide, in co-operation with other departments, the extent to which the recommendations should become a requirement of the Government of Canada. It is worth stressing that recommendations developed by a panel are advisory to government and to the regulatory agencies.

As noted above, the AECB's automatic referral list included new uranium mining facilities. In 1991, the AECB referred the six Saskatchewan uranium mining proposals for independent panel review, through the then Minister of Energy, Mines and Resources (now Natural Resources). The referral of these proposals to EARP was done as early as possible and was based on the proponents' letters of intent, not on actual licence applications.

The role of the Federal Environmental Assessment Review Office (FEARO)

Until the promulgation of CEAA, FEARO was responsible to the Minister of the Environment for the administration of the EARPGO, and for advising departments on their EA responsibilities. Once a proposal was referred for public review by an independent panel, it was the responsibility of FEARO, in consultation with the department or agency making the referral, to draft terms of reference for approval by the Minister of the Environment, to identify potential panel members, and to ensure procedures were in place to conduct the panel review. When the panel completed its work, FEARO transmitted the panels report to the Minister of the Environment and the Minister responsible for the department or agency that made the referral. In preparing the government response, FEARO typically took the lead in ensuring that the government respected the integrity of the process by giving the panel recommendations due consideration.

The role of a public environmental assessment panel

Environmental assessment panels are made up of unbiased individuals who are appointed by the Minister of the Environment to examine the environmental effects of the proposal and the directly related social impacts of those effects. With the approval of the Minister of the Environment and the Minister responsible for the initiating department, panels may also consider the general socio-economic effects of, and the technology and need for, new proposals. A panel is empowered to issue guidelines to the proponent for the preparation of an EIS, and to conduct public hearings on the project. It is the responsibility of a panel to prepare a report containing its conclusions and recommendations for decisions by the appropriate authorities. These recommendations result from a consideration by the panel of all the information submitted to it, including the public concerns the panel has heard.

The role of federal ministers

Generally under the EARPGO, it is the role of the Minister responsible for a department, board or agency to refer to the Minister of the Environment for a public review by an independent panel any project with potentially significant adverse environmental effects (or projects where public concern is such that a public review is desirable). The Minister of the Environment, as Minister responsible for FEARO, is then responsible for appointing the panel and issuing the terms of reference for the review.

Responsibility for making the panel report public rests with both the Minister of the Environment and the Minister of the responsible department. The decisions arising from panel reports are to be made by all Ministers with jurisdiction, including the Minister of the Environment. In making their decisions, the responsible authorities, including Ministers with jurisdiction, must consider the panel report, but are free to consider other sources of information and to make different value judgements.

Proposals regulated by the AECB are special cases in that many of the recommendations of panels fall within the jurisdiction of the Board, which has considerable independence as a regulatory body. Although the Minister of Natural Resources has the power to issue special directives to the Board, this power is normally reserved for extraordinary circumstances. The distinction between recommendations directed to the AECB and those directed to the government is based on whether the recommendations are related to potential licensing conditions.

The role of the Atomic Energy Control Board (AECB)

The AECB, an independent federal regulatory body established to control and supervise the development, application, and use of atomic energy, reports to Parliament through the Minister of Natural Resources. As noted, one of the roles of the AECB is to regulate the mining of uranium to ensure that the activity does not pose undue risk to health, safety, security and the environment.

Licence applications to the AECB for a project that has undergone a public review by a panel must also undergo a further internal review by AECB staff with contributions from other government agencies such as the federal departments of Environment, Human Resources Development, and Fisheries and Oceans, and their provincial counterparts. These agencies make up what is called the Joint Review Group, and this consultative process is known as the Joint Regulatory Process. All relevant comments from the Joint Review Group, and those from the public review panel, are then reflected in a recommendation from AECB staff concerning appropriate licensing action to the five-member Atomic Energy Control Board. If the Board is satisfied that the proposal is acceptable with respect to health, safety, security and protection of the environment, a licence is issued.

Case histories – projects progressing through EARP

In April 1991, six uranium mining projects in Saskatchewan were referred by the AECB for public review. Among the first projects to be referred for EARP review amid the controversy surrounding the above-mentioned Federal Court of Appeal decision, these six proposals entered the process at a very sensitive time. The five projects examined by the Joint Panel were referred under Section 12 of the EARPGO, while the sixth project, examined by the Federal Panel, was referred under Section 13 of the EARPGO. By late 1991, specific ToR had been released for both Panels and the respective reviews had begun.

The Joint Panel commenced public hearings for the first three of five uranium mining projects on 22 March 1993, and completed this phase of the review on 20 May 1993. The Federal Panel also commenced public hearings early in 1993. The marathon of hearings for the Joint Panel alone covered 20 days, beginning in Regina and ending in Saskatoon, with six northern communities in between. Sessions averaged 10 hours a day, but often lasted 12 hours. Some 300 petitioners presented opinions and positions in public meetings averaging 80 attendees. To assist intervenors in preparing briefs to present at both panel hearings, some USD 500 000 was made available by the federal and provincial governments. Both levels of government presented formal briefs at the Joint Panel hearings in support of the three projects at the opening session in Regina, and provided more supporting documentation at the closing sessions in Saskatoon. The environmental assessment of the four uranium projects reviewed to date is estimated to have cost close to USD 3 million.

The public hearings phase ended with less apprehension about the process than at the outset, but the outcome of these reviews was not to be known until the panels released their reports in late 1993.

It is also worth noting that while both panels initiated their respective reviews in early 1993, the EISs for the *Cigar Lake* and *McArthur River* projects had not been finalised and thus the review of these two projects was postponed to a later date. As well, in 1992, the Joint Panel had been asked by government to review an Underground Exploration Programme (UEP) for the *McArthur River* project, which was proposed in order to obtain data needed to prepare an EIS for the overall project. The Panel reported in early 1993, recommending that the UEP be allowed to proceed subject to certain conditions. Both governments agreed and work got underway as soon as possible.

Panel recommendations

In late 1993, the Joint Panel recommended that *DJX* proceed, subject to conditions, that *MJV* not proceed as designed, and that the *McClean Lake* project be subject to a five-year delay. The Federal-only Panel recommended that full-production underground mining of the *Eagle Point* orebody be approved, subject to certain conditions, but that approval be withheld for open-pit mining of the *Collins Bay A and D* orebodies until additional technical information on waste-rock management and decommissioning is provided by the proponents.

In early 1994, Cogema announced that it had decided to modify its plans for developing the *DJX* project at Cluff Lake, and submitted revisions to the government regulatory authorities. The revised three-phase mining plan would not require the damming and partial draining of the north end of Cluff Lake, but would require Cogema to access deeper portions of the *DJX* orebody by underground means after an initial phase of open-pit mining.

Governments' reaction and response

The federal and provincial governments responded to the recommendations of the Joint Panel on 23 December 1993. Both governments agreed that *DJX* should proceed as submitted, subject to the AECB licensing process, that *MJV* presented potential risks and should not proceed as presented, and that the *McClean Lake* project should proceed subject to the AECB normal licensing process. Governments concluded that the AECB licensing process would allow all of the technical issues raised by the Joint Panel to be considered within the context of a licence application, and provide sufficient time for the proponents to address them before the *McClean Lake* project comes into operation.

Cogema's proposed modifications were viewed by the AECB as presenting environmental impacts that were less than those predicted for the initial project and, as such, could therefore be adequately controlled. Nonetheless, the AECB invited public comment on the proposed modifications to the mining method at *DJX* to ensure that there was no significant public concern regarding Cogema's application. After receiving only minor comments, the AECB concluded that the project could proceed as re-submitted.

In March 1994, the federal government agreed with the Panel that underground mining at *Eagle Point* should proceed, subject to the AECB licensing process, but opined that open-pit mining at *Collins Bay A & D* may also be able to proceed, subject to the AECB licensing process. The AECB process would address the conditions recommended by the Panel during the evaluation of the licence applications, and would require the provision of adequate information on waste-rock management and decommissioning, as recommended by the Panel.

On 29 July 1994, the AECB referred the proposal for a redesigned *Midwest Joint Venture* to the Minister of the Environment for public review. The project is expected to be reviewed by the existing Joint Panel at the same time as the *Cigar Lake* project. ToR were prepared in close consultation with Saskatchewan's Department of Environment and Resource Management and FEARO. On 9 November 1994, the revised *MJV* uranium mining proposal was referred by the federal and provincial environment ministers for review by the Joint Panel. The EISs for *MJV* and *Cigar Lake* are expected to be submitted in the summer of 1995 in the hope that the public review process could begin as soon as possible.

Canada's new legislation - the Canadian Environmental Assessment Act (CEAA)

In June 1990, the Government of Canada introduced Bill C-78, the Canadian Environmental Assessment Act (CEAA), as a comprehensive reform of EARP. At the time, the government believed that the proposed legislation would have a greater impact on sensitising decision-making and decision-makers to the needs of the environment than any other legislation then existing elsewhere in the world. The government intended that the EARPGO would remain in force until the new legislation was promulgated, but Bill C-78 died when Parliament was prorogued in early May 1991.

The legislation, reintroduced as Bill C-13 in the new Parliament later in May 1991, had numerous amendments made that required further comments from representatives of environmental groups and industry. Meanwhile, subsequent appeals to the aforementioned court decisions resulted in the Supreme Court of Canada upholding the constitutionality of the EARPGO in January 1992, rendering compliance with the Order by all federal decision-makers a mandatory requirement. Draft regulations required by Bill C-13 were reviewed across Canada and the Bill received approval in the House in March and in the Senate in June; it received Royal Assent on 23 June 1992. The necessary operating regulations and procedures took considerable time to be finalised, and it was not until 19 January 1995, that CEAA was proclaimed by Order in Council. Projects will henceforth be reviewed under CEAA, although as noted above the Saskatchewan uranium projects will proceed under the EARPGO. The Canadian Environmental Assessment Agency, formerly known as FEARO, will oversee the new environmental review process.

The new CEAA legislation replaces the old EARP Guidelines Order, and sets out, for the first time in legislation, the responsibilities and procedures for the EA of projects involving the federal government. It reduces legal uncertainties and the need for court interpretations, sets out a streamlined process, and establishes sustainable development as a fundamental objective of the federal process. CEAA provides an environmental process for new projects, and a participant funding programme that supports public participation.

The reforms will help ensure that environmental considerations are integrated into federal decision-making processes, and will help develop greater harmonisation of EA systems across Canada. By introducing a degree of certainty in the process, the reforms will also reduce costs and time demands for all participants.

CEAA applies to projects for which the Government of Canada holds the decision-making authority whether as a proponent, land administrator, source of funding, or regulator (for listed statutes). The new CEAA process is similar to EARP in many respects, but several important changes have been introduced. These include the following: the definitions of a "project" and an "environmental effect", the introduction of "comprehensive study" and "mediation" as new EA tracks that a project may follow, the requirement to keep an ongoing record of all documents related to the assessment in a public registry, the requirement to consider the need for a follow-up programme, and the mandatory public input into specified EA tracks.

CEAA has four stated objectives, namely: i) ensure that the environmental effects of projects receive careful consideration before responsible authorities take action, ii) to encourage those authorities to take actions that promote sustainable development, iii) to ensure that projects carried out in Canada or on federal lands do not cause significant adverse environmental effects outside the jurisdictions in which the projects are carried out, and iv) to ensure that there is an opportunity for public participation in the EA process.

In addition, four guiding principles are to be followed in applying CEAA.

Early Application: the process should be applied as early in the planning stage of the project as practicable, and before irrevocable decisions are made.

Accountability: the self-assessment of projects for environmental effects by federal departments and bodies is a cornerstone of the process. A Cabinet decision by Order in Council is required in response to the recommendations of a panel report.

Efficient and Cost Effective: each project should undergo only one EA, and the level of effort required to undertake an assessment for the project should match the scale of the likely environmental effects of the project.

Open and Participatory: Public participation is an important element of an open and balanced EA process. A participant funding programme exists to promote effective public participation in the process.

Conclusion – Canada's environmental assessment process as a model

The EA process in Canada has evolved into a fairly complex set of procedures involving all levels of government, the public and the proponents. As the process matures, it will be continuously improved, with the benefit of input from all players. The difficulties experienced to date will be resolved, but new problems will undoubtedly surface that require novel solutions. Canada's EA process is truly a dynamic one.

Today, EA has become an integral part of the design process of new projects. It is now part of sound engineering practice. Indeed, obtaining financing for new projects depends on minimising environmental liabilities. The trend is established and will continue.

When the development of several new uranium mining projects in Saskatchewan was proposed, both the federal and provincial governments and the uranium industry expressed concern about the potential impacts of what appeared to be a rather burdensome public environmental review process. However, the successful advancement of three such proposals through the process has alleviated much of this uncertainty and has confirmed that all of Canada's uranium-producing operations can meet high environmental, health, and safety standards.

In Canada, the impact of the EA process on new uranium mining developments has been profound. It has changed the way projects are designed and will change the way they are brought on stream. The process has been time consuming and often difficult, but it has revealed that these new uranium mining proposals are environmentally sound and can remain so over their lifespan.

The Government of Canada is convinced that the EA process has helped to demonstrate that new uranium mining projects are being developed in a responsible manner, after full consideration has been given to the potential impacts and public concerns associated with these facilities. The lessons learned in Canada could well prove to be useful in other jurisdictions where considerations are being given to the development of new uranium production projects. However, it is understood that all processes can be improved, and both levels of government in Canada will be working closely with the uranium mining industry to make improvements in the EA process wherever possible to reduce the time and cost of reviews, while meeting the fundamental objectives.

The new global appreciation of EA may also have an impact in the marketplace. A recent motion put forward in the Swedish parliament would require uranium buyers to pay for the environmental damage caused by uranium mining in other countries. This action confirms the growing awareness that environmental considerations could have an unexpected impact on world market opportunities. Canada's exemplary EA process could give Canadian uranium producers an edge in the future.

The real impact of Canada's EA process might be that the "Canada model" is applied beyond national borders. As the world's leading supplier of uranium, Canada has attracted much attention in areas such as uranium exploration and mine development technology. If that focus is extended to Canada's positive experience with EA, the wealth of knowledge gained in Canada could well prove useful in other jurisdictions considering the development of new uranium production projects.

UPDATE ON POLICIES AND PROGRAMMES AFFECTING RADIOACTIVE WASTE MANAGEMENT IN CANADA

Uranium industry highlights

Overview

The outlook for Canada's uranium industry improved in 1996. Various factors limited the availability of uranium, and spot prices rose as uranium consumers turned to primary producers for increasingly larger amounts of uranium. While the spot price increase stalled late in 1996 the prospects for Canada's uranium industry remained favourable going into 1997.

Development work proceeds on schedule at the McClean Lake project, which will become Canada's first new uranium-producing operation since the Key Lake mine began production in 1982. If approved, the Cigar Lake and Midwest projects will also ship ore to the McClean Lake mill, where combined annual uranium production could exceed 9 000 tU. The McArthur River project, which has recently been approved, will provide ore for the Key Lake operation, greatly extending its useful life.

Uranium resource assessment

Estimates of Canada's known uranium resources have been increased sharply relating primarily to the successful underground exploration programme at McArthur River. As less than a fifth of the mineralised structure at McArthur River has been drilled in detail from underground, the potential for expanding total resources is considered excellent. As of 1 January 1996, known resources of economic interest in Canada totalled 490 000 metric tons (t) of uranium. With the closure of the Stanleigh mine at Elliot Lake, Ontario in mid-1996, known resources were reduced to 430 000 metric tons of uranium as of 1 January 1997.

Federal environmental assessment reviews

In 1996, the *Joint Federal-Provincial Environmental Assessment Panel on Uranium Mining Developments in Northern Saskatchewan* requested more information on the Midwest, Cigar Lake and McArthur River projects. The Panel reconvened and conducted public hearings for all three projects during the last half of 1996. Additional information was again requested of the proponents relating to a new tailings disposal plan for use at the McClean project. While the Panel announced that insufficient data had been provided to proceed, it is hoped that the Cigar Lake and Midwest proponents could supply this necessary information so hearings might be concluded in early 1997. The Panel completed its review of the McArthur River project by year end 1996, and commenced preparing recommendations to governments. Its report was submitted to governments early in 1997, a response to the Panel's McArthur River report was released before mid-year by governments, approving the project.

Other developments

On 30 June 1996, Rio Algom Limited closed its Stanleigh operation at Elliot Lake, Ontario, as planned, winding up 40 years of uranium production in Canada. During the first half of 1996, Rio Algom reportedly shipped more than 400 tU from the Stanleigh operation to Ontario Hydro.

With construction of Cameco Corporation's new CAD 10 million recycling facility at Blind River, Ontario, completed in June 1995, commissioning proceeded during 1996. Using an innovative process to convert by-product liquid raffinates into a dry powder form, the facility will reduce volumes by 75%. These solids will be stored on site prior to shipment to a facility for final recovery of the remaining uranium. Cameco developed this process to reduce by-product volumes because of the closure of the Stanleigh operation at Elliot Lake, north of Blind River.

Cameco conducted a pilot study in 1996 to determine the feasibility of recovering nickel and cobalt from Key Lake tailings. If proven to be economic, a CAD 45 million extraction plant, capable of handling over 800 tonnes of tailings daily, could be in operation as early as 1998. The reprocessing of several million tonnes of tailings from the original surface impoundment would take a decade or more, result in production of 3 175 tonnes of nickel and 227 tonnes of cobalt annually, and permit these tailings residues to be re-deposited in the new Deilmann pit subsurface tailings facility. While extraction proved to be technically feasible, the market outlook for cobalt and nickel by early 1997 led to a postponement of the commercial development plan.

Radioactive waste management

Policy framework for radioactive waste

In July 1996, the Minister of Natural Resources Canada announced a Policy Framework for Radioactive Waste that guides Canada's approach to the disposal of nuclear fuel waste, low-level radioactive wastes and uranium mine waste in Canada. The Framework lays out the ground rules and sets the stage for the further development of institutional and financial arrangements to implement disposal in a safe, environmentally sound, comprehensive, cost-effective and integrated manner. The federal government has the responsibility to develop policy, to regulate, and to oversee radioactive waste producers and owners in order that they meet their operational and funding responsibilities in accordance with approved disposal plans. The Framework recognises that there will be variations in approach in arrangements for the different waste types.

Uranium mine and mill tailings

In Canada, close to 200 million tonnes of uranium mine wastes and mill tailings have been generated since the mid-1950s. These comprise about 2% of all mine and mill tailings in the country. Most of the existing uranium tailings are located in the provinces of Ontario and Saskatchewan. There are twenty-two tailings sites, nineteen of which are no longer receiving waste material. Only the operations in Saskatchewan are now active.

With regard to financial responsibility for decommissioning and long-term maintenance of the tailings, the general policy in Canada is that the producer pays. The AECB requires that operators provide financial assurances that decommissioning of uranium facilities will take place in a responsible and orderly manner in the short and long term. Where a producer or owner is unable to pay, responsibility for decommissioning would rest with the Canadian federal and provincial governments.

In January 1996, a Memorandum of Agreement (MOA) on cost-sharing of management of abandoned uranium mine tailings was signed between the federal and Ontario governments. Canada and Ontario had been discussing the issue of financial and management responsibilities for abandoned uranium mine tailings since 1984. The MOA recognises that present and past producers of uranium are responsible for all financial aspects of the decommissioning and long-term maintenance of uranium mine sites, including the tailings. In the case of abandoned sites, the MOA outlines how governments will share the long-term management responsibilities and associated costs.

Federal environmental assessment reviews

The decommissioning plans for four uranium mine and mill tailings sites in the Elliot Lake area of Ontario were referred by the AECB to the Minister of the Environment for a federal environmental assessment and review. It will be the first time that the decommissioning of uranium tailings will have gone through this process. The three-member *Elliot Lake Environmental Assessment Panel* conducting the review provided recommendations to the federal government in June 1996. The government is currently considering these recommendations and hopes to respond by spring 1997.

A review of decommissioning plans for another uranium mine site in the Elliot Lake region was initiated in late 1996. This review will be conducted under the requirements of the new Canadian Environmental Assessment Act which came into force in 1995.

• China •

BASIC FACTS OF URANIUM MINING AND MILLING

Uranium Mining and Milling was initiated in China in 1956. A complete system of uranium mining and milling, including research, design, construction and operation, was established over 40 years ago. The mines and mills are located in 14 provinces.

Characteristics of uranium resources

Small deposits; diversity of ores; low grade ores.

Table 1. **Waste produced during uranium mining and milling**

	Type	Radon release Bq/t ore	Aquifer effluent t ore	Residue t ore	
				Rock	Tailings
Uranium mines	Underground	7.01×10^3	0.3~8.0	0.7~1.5	
	Open-pit		0.1~0,6	5~8	
Milling complex	Sorting	2.0×10^1	0.5~1.0		0.2~0.3
	Milling	5.1×10^2	8.0~10.0		~1.2

Table 2. **Annual average content of nuclides in effluent from several uranium mines and mills over 30 years (Bq)**

Nuclides		U	^{230}Th	^{226}Ra	^{222}Rn	^{210}Po	^{210}Pb
Mines	gaseous	1.37×10^8	6.86×10^7	9.68×10^{13}		7.25×10^7	6.95×10^7
	liquid	1.37×10^{11}	1.65×10^9	1.81×10^{10}		1.25×10^8	3.48×10^8
Mills	gaseous	2.90×10^9	3.25×10^7	3.25×10^7	5.86×10^{12}	3.25×10^7	3.25×10^7
	liquid	5.66×10^9	5.87×10^8	2.59×10^{10}		1.46×10^9	3.12×10^9
Complex	gaseous	6.00×10^8	3.01×10^8	3.01×10^7	1.78×10^{14}	3.01×10^8	3.00×10^8
	liquid	6.48×10^{10}	1.10×10^{10}	3.11×10^{10}		9.31×10^9	1.64×10^{10}
TOTAL	gaseous	3.58×10^9	3.68×10^8	3.94×10^8	2.76×10^{14}	3.72×10^8	3.68×10^8
	liquid	1.91×10^{11}	1.22×10^{10}	7.07×10^{10}		1.01×10^{10}	1.85×10^{10}

Table 2 shows that ^{222}Rn represents the largest quantity of gaseous radioactive effluent, while other nuclides are comparatively insignificant. In aquatic dispersal of radionuclides from uranium mining and milling, the major nuclide is uranium, while the others (^{226}Ra, ^{230}Th, ^{210}Po and ^{210}Pb) tend to be somewhat less abundant. At the mills, ^{226}Ra is the major nuclide in the effluent.

The distribution of annual individual effective dose equivalents to critical groups from uranium mining and milling is given in Table 3.

Table 3. **Distribution of annual individual effective dose equivalents to critical groups from uranium mining and milling**

mSv/year	Dose distribution (%)			
	<0.1	>0.1, <0.25	>0.25, <1	>1, <2
Mine	21.2	13.2	9.2	4.0
Milling plant	2.2	1.1	9.9	4.0
Complex	13.9	7.7	14.7	1.1
Total	37.3	22.0	33.8	6.9

It can be seen from Table 3 that the dose equivalents to critical groups below 1 mSv represent 93.1% of total facilities-years; those lower than 0.25 mSv equal 59.3%; and those lower than 0.1 mSv equal 37.3%. The maximum dose equivalent is 1.82 mSv.

Table 4. **Distribution of nuclides by dose equivalent from gaseous and liquid effluent to critical groups (man.Sv)**

Nuclides	^{222}Rn	U	^{230}Th	^{226}Ra	^{210}Po	^{210}Pb	Total	Per cent
Gaseous	1.27×10^{1}	8.85×10^{-1}	5.5×10^{-2}	2.75×10^{-2}	1.34×10^{-2}	1.15×10^{-1}	13.80	76.7
Liquid		$2.33\text{-}10^{-1}$	1.35×10^{-2}	9.48×10^{-1}	5.46×10^{-2}	2.94	4.19	23.3
Total	1.27×10^{1}	1.12	6.85×10^{-2}	9.76×10^{-1}	6.82×10^{-2}	3.06	17.99	100.0
Per cent	70.6	6.2	0.4	5.4	0.4	17.0		

Table 4 is a list of collective effective dose equivalent for each nuclide distributed to critical groups, together with their relative fraction, resulting from gaseous and liquid radioactive effluents from uranium mining and milling. Of the nuclides, ^{222}Rn is the major contributor to collective effective dose with the equivalent fraction of 70.6%. The nuclides ^{210}Pb, U, ^{226}Ra, ^{230}Th and ^{210}Po, listed in decreasing order of abundance, with a respective relative fraction of 17.6, 6.2, 5.4, 0.4 and 0.4%.

Environmental issues in uranium mining and milling

After 40 years of development, many uranium mines and mills have been closed. Decommissioning and remediation of the sites have reached different stages of completion. Several problems related to reclamation activities remain to be addressed, including:

- The measurement of annual average radon flux and the thickness and pattern of cover material.

- The management of liquid mine effluent and the assessment of the impact of water from tailing impoundments on underground water.

- The safety analysis of tailing dams and reinforcement method.

• Czech Republic •

INTRODUCTION

The state company DIAMO is the only mining and ore processing organisation in the Czech Republic involved in the extraction and processing of uranium ores. It is a 100% state-owned organisation managed by the Ministry of Industry and Commerce of the Czech Republic. Today, DIAMO's main programme includes continuing uranium production at decreasing levels and the decommissioning and restoration of uranium production sites.

Environmental impacts of extraction and processing activities of DIAMO

The impacts of released mine and waste water, and of waste management on the environment

The state of the environment in the regions where DIAMO has activities, is monitored under a programme which was approved by the administration of the state inspection. The results of the monitoring programme are regularly reported to the state authorities.

Emissions to the atmosphere

Emissions from boiler plants are the only emissions to air that are monitored. With the exception of the Dolní Rozínka uranium production centre all emissions from thecnical installations are very low, and are nearly at the detection limit. The values for ammonia are somewhat higher at Dolní Rozínka.

Release of mine and waste water

The overall amount of water released in the critical regions (i.e. Stráz, Dolní Rozínka, Príbram) is monitored continuously.

Waste management

In 1995, waste management was in compliance with the approved programmes. The continuing decommissioning activities resulted in an enhanced production of wastes of the category Z, N (wastes contaminated by radionuclides). The volume of other wastes produced begins to be affected by the transition to the decommissioning regime.

Underground water in the deposit Stráz affected by the chemical extraction of uranium

In the district of Northern Bohemia the mining activity of DIAMO is concentrated in the eastern part of the district of Ceská Lípa, in the vicinity of Hamr n. Jezere and Stráz pod Ralskem. In this location, at the end of the 1960s, two mining methods – the classical deep mining and the method of underground chemical leaching in situ (ISL) in the ore layer (chemical extraction) were forced to be developed in parallel. The in situ procedure of chemical extraction leached uranium in a closed cycle using sulphuric acid at a concentration of 2 to 5%. The acid solution was introduced directly into the ore layer using a grid of technological wells.

In the past, the means chosen for uranium mining in Northern Bohemia proved to be inappropriate. The extensive development of both the conventional and in situ extraction methods in a relatively small area resulted in highly unfavourable mutual influences. Moreover, this development was not accompanied by the introduction of appropriate measures and technologies for minimising the environmental impacts of the extraction technology.

The ISL mining was extensively applied for more than 20 years in an area exceeding 600 hectares. The ultimate result of this strategy was the contamination of 186 million m^3 of groundwater in an area of 24 km^2. One part of the contaminated area is located outside of the limits of the leaching fields.

Overview of the contamination of the Cenomanian (ore bearing) complex

Contaminated volume of 186 million m^3 in an area of approximately 24 km^2 include:

Principal contaminants	
SO_4^{2-}	3 792 x 10^3 t
free H_2SO_4	984.2 x 10^3 t
NH_4^+	91.4 x 10^3 t
Al	413.0 x 10^3 t
U	~1.0 x 10^3 t
Total dissolved solids (TDS)	4.8 x 10^3 t

Overview of the contamination of the Turonian horizon

The amount of contaminated water is approximately 80 million m^3 in an area of 7.5 km^2.

Principal contaminants	
SO_4^{2-}	22.0 x 10^3 t
NH_4^+	1.3 x 10^3 t
Total dissolved solids (TDS)	25-30 x 10^3 t

The ore-bearing horizons and zones of active leaching are situated below the Upper Turonian horizon, which contains large resources of potable water. The use of acid ISL extraction technology, together with insufficient casing of the technologic wells and, also possibly the insufficient technologic discipline, resulted in a secondary contamination of the overlying Turonian aquifer.

Target of the remediation

Extensive exploration and evaluation programmes were conducted from 1992 to 1994 in the Stráz area. The extent and direction of this work was determined by the Decision of the Czech Government No. 366 of 20 May 1992. Results of this work was used for modelling different remediation options for the deposit. Based on these results, Decision of the Czech Government No. 170 of 6 March 1996 directed that the uranium mining operation change from the chemical extraction to the remediation regime. The preparatory period of the remediation regime was then initiated. The construction of an evaporating plant that could facilitate the release of 6 m³/minute of purified water into the surface watercourse was then completed. This treatment is necessary to stabilise the spatial extent of the contamination. In the remediation stage it is planned to construct facilities to extract the soluble salts from the deposit. Until then it will be necessary to introduce the concentrated solutions from the evaporation plant back into the central part of the deposit.

The objectives of the remediation in the deposit following ISL extraction are to:

- Gradually lower the content of the dissolved solids in the Cenomanian aquifer to a limit that will restrict the risk of endangering the overlying Turonian aquifer, as this is an important resource of potable water.

- Gradually lower the content of dissolved solids in the Turonian aquifer to the level of the Drinking Water Quality limits for individual substances according to the Czech standard CSN 75 7111.

- Gradually integrate the ground surface area over the leaching fields into regional systems of ecological stability.

The following table gives an overview of the impoundments and waste rock dumps operated by DIAMO.

Locality	Beginning of operation	End of operation	Total area (hectares)	Volume of deposited material ($10^3 m^3$)
Stráz mining area				
Stráz p.R. – I. Stage	1980	1992	93.5	9 236.4
Stráz p.R. – II. Stage	1991	1992	93.5	800.0
TOTAL			**187**	**10 036.4**
Príbram mining area				
Dubno, KI	1988		33.0	86.8
Dubno, KII	1980	1988	11.1	152.5
TOTAL			**44.1**	**239.3**
Jáchymov mining area				
Jáchymov – Nejdek			9.6	1 440.0
Jáchymov – Elías			10.5	1 358.0
TOTAL			**20.1**	**2 798.0**

Locality	Beginning of operation	End of operation	Total area (hectares)	Volume of deposited material ($10^3 m^3$)
Rozná mining area				
Rozná, K-I	1968		62.75	8 836.7
Rozná, K-II Zlatkov	1980		27.45	990.8
TOTAL			**90.20**	**9 827.5**
Mydlovary (Mape), tailings area				
K-I	1962	1984	24.40	5 551.0
K-II	1967	1981	76.62	10.3
K-II	1980	1985	33.74	4 354.0
K-IV/CIZ	1988	1989	35.71	403.0
K-IV/C2	1985	1988	31.92	1 708.0
K-IV/D	1985	1991	35.76	661.0
K-IV/R	1984	1991	31.02	950.0
K-IV/CF			23.35	0
TOTAL			**293.52**	**13 637.3**

Tailings impoundments and waste dumps

The tailings impoundments of the CHÚ s.p. represent one of the most complicated problems faced by DIAMO. This exists both from the operational point of view, and moreover, regarding their close-out and recultivation The most important problems to be addressed are:

- The "overbalance" or excess water, and the search for the optimal water purification technology.
- Purification of seepage water.
- Design and construction of a cover for the tailings impoundment.
- Integration of the recultivated area into the surrounding landscape.

The remediation of waste rock dumps, resulting from the extraction of ore from conventional underground mines will be solved in the following ways:

- Use of the waste rock material for the follow-up remediation of the impoundments and their definite decommissioning and releasing the area for further use.
- Recultivation of the dumps and their integration into the natural environment.

Principal decommissioning and remediation activities of DIAMO

The remediation of the tailings from uranium processing is required at all of the subsidiaries following the suspension of the uranium activities in the Czech Republic. Since 1993, all remediation and decommissioning measures are funded from the state budget of the Czech Republic.

The remediation and decommissioning measures are controlled from DIAMO's headquarters which secures the required funding from the state budget. The remediation and decommissioning activities are carried out by the DIAMO subsidiary responsible for the area in which the facility is situated.

The ongoing remediation and decommissioning activities are summarised below.

In the region of competence of Túu Stráz pod Ralskem

Decommissioning of the Hamr mine

The decommissioning project for the Hamr mine has been prepared and is approved. The project involves backfilling of the underground sites. The annual volume of backfilling is about 100 000 m^3. The backfilling will be finished in 2001.

Decommissioning of the Stráz chemical processing plant

The technologic equipment is being decommissioned and the Stráz chemical processing plant buildings will be decontaminated. It is proposed to release the decontaminated buildings for further use.

Remediation of the Stráz impoundments

The conceptual design for the remediation of the Stráz impoundments has been prepared. The first step involves liquidation of the free water, to be followed by transferring the tailings in the stage II impoundment, together with the contaminated material from the dam, into the stage I basin.

In the region of competence of the subsidiary Cht Stráz pod Ralskem

As of 1 May 1996, the preparatory phase of remediation of the Stráz ISL well field was put into operation. This requires both evaporating about 6.5 m^3 per minute of solutions, and the release of about 6 m^3 per minute of distillate into the surface watercourse. The operation of the evaporating plant started on 1 October 1996. Another important project involves the plugging of old exploration drillings. This measure is necessary to prevent the release of leaching solutions from the leach zone through the old wells into the overlying Turonian aquifer.

In the region of competence of the subsidiary Geam at Dolní Rozínka

Decommissioning of the Olsí mine

Continuing work at the Olsí mine involves recultivation of the dumps and plant area, as well as the continuation of mine water processing. Processing of the mine water is required as it contains high concentrations of uranium, iron and manganese.

Decommissioning of the Jasenice-Pucov mine

Recultivation of the Jasenice-Pucov mine is finished. The processing of mine water, involving the removal of uranium and radium, is continuing at the site.

Decommissioning of the Licomeríce-Brezinka mine

At present recultivation of the Licomeríce-Brezinka mine area is projected. Mine water processing continues at the site to remove uranium, radium, iron, and manganese. A characteristic of the site is the continuing biologic leaching of uranium in the mine shaft.

In the region of competence of the SUL Príbram subsidiary

The subsidiary SUL is in charge with the decommissioning and remediation activities in several regions: Príbram, Western Bohemia, Horní Slavkov, Jáchymov, MAPE Mydlovary, and Okrouhlà Radoun. It also administers the sites at which mining activities were discontinued in the past: the region of Jáchymow, Krkonese Mountains, Zálesi u Javorníka and others. In these areas an inspection of the monitoring systems and the maintenance of shafts and surface mine workings are conducted twice a year.

The most important remediation and decommissioning activities are described as follows:

Remediation of impoundment II – Bytíz Príbram

The technical recultivation of the Bytíz Príbram impoundment II has been completed. This includes implementing steps to reduce gamma dose rates at the impoundment surface, as well as preventing blown dust formation. The topography of the surface was also adjusted to the final shape.

Processing of the impoundment water at Príbram

The preparatory stage for remediation of the impoundment No. I at Príbram, involving reduction of the free water volume, is underway.

Recultivation of the Jeroným Abertamy waste dump – region of Jáchymov

The recultivation of the Jeroným Abertamy waste dump had been damaged by the uncontrolled removal of waste rock. Originally the dump was recultivated allowing natural vegetation to be re-established on the site.

Recultivation of the Vítkov II mine – Western Bohemia

At the Vítkov II mine, technical recultivation and also a partial biological recultivation has been carried out. Water processing is continuing at the site.

Processing of mine water at the Zadní Chodov Mine

At the Zadní Chodov Mine site an uncontrolled flow of mine water occurred below the projected mine water discharge point. It was therefore necessary to take immediate measures to prevent contamination of surface water with radionuclides. The content of uranium in the mine water reaches 15 to 20 mg/litre.

Recultivation and processing of mine water at Okrouhlá Radoun

The recultivation of the waste rock dump and the pumping and processing of mine water for the removal of radionuclides continues at the Okrouhlá Radoun mine.

Construction of a mine water processing plant at Horní Slavkov

One of DIAMO's greatest remediation projects is in operation at Horní Slavkov. Mine water will be processed from the abandoned shafts. Radionuclides, Fe, Mn and some other elements are being removed to decontaminate the water. The need to realise this construction emerged from the stocktaking of old liabilities.

Recultivation of the tailings impoundments of MAPE Mydlovary

The most extensive remediation project of DIAMO is recultivation of the tailings impoundments of MAPE Mydlovary. The duration of this project is estimated to take several tens of years. At present a study is being prepared for the selection of the optimal remediation concept. At the same time several options are being evaluated using pilot projects. The seepage water in the area of the impoundments is being processed at the rate of 6 to 7 m^3/hour.

Conclusion

Mining and processing of uranium ores in the Czech Republic led to major environmental impacts. The mitigation of the impacts will require a long-lasting remediation procedure. It will probably continue for many years after the year 2000 and will require considerable financial resources.

Today, in parallel with the continuing reduction of the level of mining of uranium ores, the remediation and decommissioning activities are becoming the main programme of DIAMO.

DIAMO is supported with funds from the Government of the Czech Republic. It is responsible for maintaining the technical quality of the remediation and decommission measures and for the effective use of the funds from the state budget.

One of the means for maximising the cost effectiveness of the remediation and decommission measures is quality assurance according to ISO 9000. This is done with the purpose of receiving a quality certificate for remediation work. This practice is also connected with the possibility of introducing a system of environmentally oriented management.

• Estonia •

HISTORY OF URANIUM PRODUCTION

The history of uranium production in Estonia is related to the Silamäe Metallurgy Plant, located in north-eastern Estonia. This plant was constructed in 1948 to process uranium bearing ores. It was first used to recover uranium from alum shale mined in Estonia until 1963, then used to process uranium ores from eastern Europe. By 1977, an estimated 4 million tonnes of uranium ores were processed at the plant. Later, the plant was used to recover niobium, tantalum and rare earths. The waste repository contains 12 million tonnes of tailings and other wastes, plus an estimated 4 million tonnes from uranium tailings, oil shale ash and waste from processing of loparite for niobium and tantalum.

ENVIRONMENTAL CONSIDERATIONS

From 1992 to 1994, an international co-operation project assisted Estonian specialists in conducting an environmental site assessment of the Sillamäe mill tailings. The results of this assessment are being used to estimate the impact on the environment and to plan site remediation and long-term closure of the uranium tailing impoundment. The findings of the investigation are summarised below.

From 1948 to 1959 the uranium tailings were heaped on the surface of the lower coastal terrace near the plant on the immediate waterfront of the Gulf of Finland. The waste repository was established in the same area in 1959. The repository has been reconstructed several times since then. During the 1969-1970 reconstruction, part of the solid uranium tailings from the repository was used as fill material to construct a higher dam. At present the repository has the form of an oval retention impoundment with an area of 330 000 m^2. The impounded material is surrounded by a dam which is 25 metres above sea level. No cover material has been placed over the impoundment.

Radon and its progeny are emitted from the uncovered repository. This is the major source of radiological impact on the population of Sillamäe. The resulting annual individual doses are of the order of 0.2 millisievert. The impact of water leaking through the repository, and from the neighbouring closed mines discharging into the Gulf of Finland is much less. The resulting impact is observable only near the repository. This discharge results in the collective committed 50-year dose of about 1 manSv, and in an individual committed effective dose of about 1 microSv. The main environmental concern defined by the international co-operation project in 1992-1994 was the stability of the repository. A potential collapse of the dam, or a landslide should not be neglected.

Environmental concerns also arise in connection with contamination present at the former transportation (railway) terminal and the storage area for imported ores located outside the plant. The dose rates are rather high at many locations in these areas. Only limited funding has been available to pay for environmental clean-up. Consequently, these areas have only been partly cleaned by the plant staff.

• Finland •

URANIUM MINES AND RELATED SITES

The radiological state of the former Lakeakallio and Paukkajanvaara mines have been monitored by the Finnish Centre for Radiation and Nuclear Safety since 1974. The pits and waste heaps were covered by soil by 1993. The other sites are very small and need no special measures, with the possible exception of Nuottijärvi.

Uraniferous phosphorite-type mineralization is associated with the Vihanti sulphide ores, and some waste heaps could exist at the former mine area. The Korsnäs lanthanide concentrate, stored at the former mine, will soon be covered by soil. Uranium-bearing veins could still be exposed at the prospecting area of the Pahtavuoma sulphide mine. Low-grade uranium-bearing mine wastes remain as a result of mining at Sokli, Talvivaara and Juomasuo. Uranium of the Sokli regolith occurs in the residual pyrochlore, and in the secondary apatite and gothite.

Uranium mines and test pits

Mine	Operation	Ore output (tons)	Produced U (tons)	Remarks
Lakeakallio	1957-1959	557	1	Pilot plant
Paukkajanvaara	1958-1961	40 325	30	Mining
Hermanni	1959	0	0	Test pit
Riutta	1959	0	0	Test pit
Luhti	1960	0	0	Test pit
Nuottijärvi	1965	867	0	Pilot plant

Uranium as a by-product

Mine	Operation	Remarks
Vihanti	1952-1992	27 Mt of sulphide ore mined partly with U
Korsnäs	1967-1971	36 000 t of lanthanide concentrate
Pahtavuoma	1974-1976	Exposed U ore outcrops possibly still open
Sokli	1978-1980	56 720 t of apatite ore mined containing U
Talvivaara	1981 -1982	37 000 t of Zn-Cu ore mined containing U
Juomasuo	1992	17 635 t of Au ore mined containing U

Other sites with possible uranium contents

- Waste heaps of gypsum at the fertiliser plants that have processed sedimentary phosphate rock contain some uranium; for instance, at Oulu, Uusikaupunki, and partly at Siilinjärvi.

- The ash from the peat or coal-burning power plants could have an elevated uranium content.

Legislation and regulations

Mining in Finland is regulated by the Mining Law of 1965 or by the Land Extraction Act of 1966. Uranium is especially tied to the Nuclear Energy Act of 1987. These laws have sections dealing with the opening, operating and shut-down of uranium mines, as well as with the disposal of nuclear wastes and spent fuel. New considerations have arisen from the Environmental Permit Procedure Act of 1992.

Research

An international research programme on radionuclide transport analogy has been started in the surroundings of a uranium deposit at Palmottu in southwestern Finland. The deposit is situated within a former prospecting target area. At the site the cored drill holes still remain open and offer suitable conditions for hydrogeological and geochemical studies. The main object of the programme is to study the environmental effects of a uranium deposit in its natural state.

• France •

REMEDIATION OF MINING SITES IN FRANCE

Legal setting

French laws and regulations have established that mineral substances which can be found in the earth do not belong to the owner of the land but to the French State. The government may grant rights on exploration and mining of orebodies to mining companies which request them.

The legal framework of exploration and mining of orebodies of mineral substances and fossil fuels considered as "mines" is governed by the *Code minier* (Mining code) which established a law dated 26 May 1955 on the consolidation of legislative texts concerning mining activities and which is a compilation of all the legislative texts on mines. This law was modified several times since 1955 and has been completed with regulations on mine policies during opening, exploitation and the closing phases of mines and with other different regulations such as the "Réglement Général des Industries Extractives (RGIE)" (General regulation on extraction industries) which defines the rules of sanitation and safety to be observed by mining companies.

In this context the *Code minier* sets up the procedure to be applied at the end of the exploration or mining of a lease and the remedying duties due by the prospecting or mining companies when the activities are stopped or when the mines are closed. The 9 May 1995 decree about mine opening and policy is by law the latest version of the *Code minier* updated as of 15 July 1994.

This 15 July 1994 law has simplified the control procedure of the stopping of mining activities and has improved its efficiency by creating a unique procedure for the closing of mining works and mining activities which can be applied both in the case of an active mining lease and in the case of a mining lease at end of validity.

In the course of definitely stopping the mining works and, as it may happen, at the end of each working stage, the company which explores or exploits a lease must provide an end-of-mining-works statement to the administration in charge of the control and supervision of mining sites. This declaration must be sent to the Préfet,[1] at least 6 months before the definite end of the mining works.

In its end-of-mining-works statement, the company specifies "the measures which will be taken to preserve the interests mentioned in the *Code minier*, in a general way to stop the various after-effects, troubles and nuisances of any kind produced by its activities and to allow, if possible, for the renewal of mining activity".

The measures proposed by the mining company must allow for restoration of the mining site and ensure protection of the environment on the whole, protection of water resources, cultural assets of a region, protection of the means of communication and of agricultural interests; they have to make sure that buildings will remain in good condition, that the staff will work in safe and healthy conditions and they must safeguard public health and safety (these interests are enumerated in the 79th article of the *Code minier*).

According to the provisions of the *Code minier*, the operator must take remediation measures to preserve the water resource. With this aim, a hydrological assessment should be made that includes:

- An appraisal of consequences of the end of mining works on water and on the uses of water.

- Indication of measures of compensation which are contemplated in order, for instance, to balance the flow decrease of some rivers which results from the diminishing of mines dewatering.

In addition, the operator needs to indicate if some parts of the mining areas will be used for other activities such as industrial ones.

The end-of-mining-works statement is examined by the administration in a way similar to the investigations linked with a request for mine opening. In particular, the local administrative authorities and the mayors of the towns and villages concerned must be consulted. This process is organised by the Regional Authorities for Research, Industry and Environment (DRIRE) and the *Préfet*'s technical support to implement the mine policy.

Once these requirements are satisfied, the *Préfet* can, in a prescribed period of 6 months after receipt of the declaration, and if he thinks it is necessary, order by decree some supplementary measures that the operator will have to complete to reclaim the site. If the measures proposed by the operator seem sufficient to him, the *Préfet* gives his consent. When the operator receives this agreement he may proceed to site reclamation. If no response is given to the operator within the prescribed time, he may proceed to site restoration in the way indicated in his declaration and the mining works are then considered to be finished.

---

1. The Préfet is the representative of the French government in an administrative district called a "Département". In order to achieve his tasks, particularly in the field of control and supervision, he is assisted by several administrative departments. In the case of industrial activities the department involved is the "Direction Régionale de l'Industrie, de la Recherche et de l'Environnement (DRIRE)" (Regional Directorate for Industry, Research and Environment).

Remediation of uranium mines in France

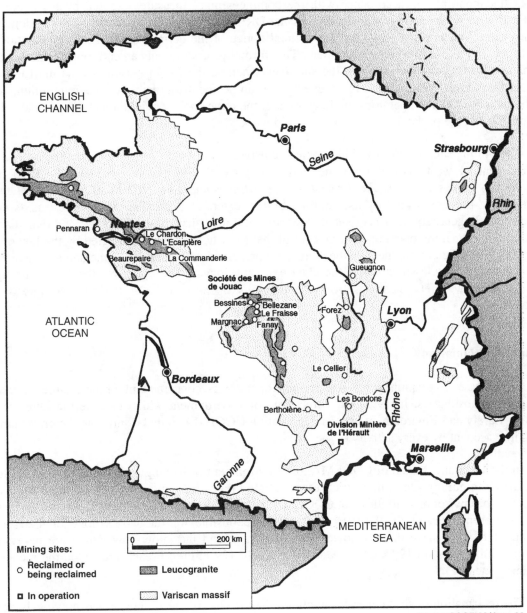

Source: **COGEMA/June 96**

When the remediation activities are completed, the *Préfet* orders a verification report to check that the works undertaken are in line with the measures foreseen by the operator in his end-of-mining-works statement, and in accordance with the complementary measures ordered by his decree. This verification report is an objective element for the DRIRE to assess the reclamation works done on the mining site. It enables the *Préfet* to give official notice of the definite stopping of remediation work and of the end of use of mining equipment. This formal procedure puts an end to the mine policy and to the administrative supervision of the site. Nevertheless, the *Préfet* may intervene until the expiry date of the mining lease in case of an event or of an accident ascribable to previous mining works. With this procedure the operator is no longer under the mines policy regulations but he remains legally responsible in case of damages caused by the old works.

The mining sites strictly speaking come exclusively under the *Code minier*, but the mills and the management of waste issued from the mills are governed by the law of 19 July 1976 on the *Installations Classées pour la Protection de l'Environnement (ICPE)* (Classified plants for environment protection) and by its modified by-law dated 21 September 1977. These texts contain some very stringent requirements related to closing activities, dismantling of plants and remediation of sites. In this frame, the operator must be careful that the whole interests protected by the law of 1976, and particularly the environment on the whole, are preserved and that none of the damages which have led to the previous classification of the plant exist any longer. The law gives the *Préfet* the right to order some new safety measures from the previous operator, even if the site has already been declassified.

Means of reclamation

In the case of uranium, only one mining operator is concerned in France, either directly or through its subsidiaries. In 1993, the Directorate of Environment which became the Directorate of Quality, Safety and Environment, was created within COGEMA to make sure that the environment is taken into account by every activity of the Group.

In the Uranium Branch of COGEMA, the Environment of Mining Sites Service co-ordinates the remediation of mining sites with a strict respect of the French regulation and beyond these rules with a respect of the environmental rules that COGEMA has itself defined. The Mining Centres which are the direct representatives of the DRIRE are legally responsible for the works done in their area of investigation. They have the manpower, the equipment and the know-how which are necessary to reclaim mining sites. In 1994, COGEMA spent nearly 129 million French Francs on the reclamation of mining sites.

The reclamation of sites depends largely on the activity to which the site was previously devoted (open pit or underground mine, mill, occurrence of waste dump and nature of this waste, etc.) and on local conditions (geographical, geological, hydrological and human environments). So it is quite difficult to define a general technical process applicable to every project. Every site must be studied on a case by case basis in order to achieve a viable reclamation project which conforms to the regulations.

The main types of sites to be reclaimed are the following: underground mines; open pit mines; heap-leached dumps; uranium mills and the tailings ponds which are connected with them; and annexed areas (building offices, storage depots, stockpile areas and barren rock dumps).

The annexed areas do not present any difficulty to be reclaimed. The equipment and buildings are dismantled, soils are cleaned up, decontaminated if needed, and the areas can generally be returned to their previous activity or assigned to new ones.

In agreement with a third party (a local community for example), the open pit areas can be converted into agricultural reservoirs, diving training-centres, fish hatcheries, wild life reservations in order to be used in the best way to match with local or regional development of the surroundings.

Prior to achieving this conversion, the necessary works to ensure safety are carried out, the slopes are smoothed, the pits are partly or totally filled with barren rocks in order to fit in with the landscape. Once a layer of soil has been be spread on to the surface, these areas can be sown with grass, and activities such as agriculture or industry can then take place.

Mining sites with waste disposals in France*

Name	Type of activities	End of activity	Reclamation stage
St. Hippolyte	Underground mine, Heap leaching	1969	finished in 1994 low grade ore disposal – radiological impact control in progress
Lachaux Rophin	Underground mine, mill, Tailings pond	1955	finished in 1985 regular supervision
Gueugnon	Mill, tailings pond	1980	finished regular supervision
Les Bois Noirs St Priest	Underground mines, Open pit, Mill, tailings pond	1980	finished regular supervision
St Pierre du Cantal	Open pit, heap leaching, mill	1985	in progress regular supervision
Le Cellier	Open pit, underground mines, heap leaching, Mill, tailings pond	1990	finished in 1991 regular supervision
Lodève	Open pit, underground mine, heap leaching, mill, tailings pond	1997	in progress several sites and disposal regular supervision
Bertholène	Open pit, underground mine, leaching in stalls	1994	in progress regular supervision
Le Bernardan Jouac	Open pit, underground mine, heap leaching, mill, tailings pond	*mine in operation*	in progress regular supervision
La Crouzille - Bessines	Open pit, underground mines, heap leaching, mill, tailings pond	mill 1993 mine 1995	in progress 4 different disposals regular supervision
La Ribière	Open pit, mill, heap leaching	1985	finished in 1992 regular supervision
La Commanderie	Open pit, Heap leaching	1976	finished regular supervision
L'Ecarpière	Underground mine, heap leaching, mill Tailings pond	mine 1990 mill 1991	in progress regular supervision

* Source: "Inventaire national des déchets radioactifs – ANDRA 1996".

The barren rocks which have been extracted during the mining activity are used for the safety works and the refilling of the pits. The remaining dumps are smoothed to blend harmoniously into the landscape, and are vegetated.

The shaft-heads and the accesses to the main drifts of the underground mines are carefully blocked and filled after the underground works have been placed in safe conditions. For example the cavities near the surface are stabilised if needed. When the mine is flooded, the water resurgences (if any) are controlled and treated (pH, heavy metals, uranium, radium) if needed, so that they can remain below the level which allows them to be cast-off into the environment.

The most substantial remediation work is certainly the one connected with the reclamation of the tailings disposal areas. The operator has to:

- Ensure the protection against external exposure.
- Limit, by various means, the dispersal of radon in the environment.
- Ensure a perennial mechanical stability of the disposal.
- Limit the leaching of tailings by groundwater and avoid the leakage of products.
- Prevent a misuse of the material.
- Recontour the disposal area in harmony with the landscape.

Some elaborate preliminary studies are made on these sites, and many outside experts are required in the fields of geology, geochemistry, mineralogy, radiology, hydrology, soil engineering and landscape.

The remediation of tailings disposal areas must first ensure the long-term stability of the disposal, to avoid any human or animal intrusion into the tailings and to limit the intrusions that could occur later on the site. It must also minimise the concerned surfaces, reduce the remaining impact on the surroundings and take the wishes of the local communities into account. Finally, if possible, it has to allow the use of the surface for some human activities.

The dams of the tailing ponds are redrawn, reinforced if needed, in order to ensure their long-term stability and are finally revegetated. After natural and progressive drying, the fine tailings which lie behind the dams are covered with coarser waste issued from heap-leaching and then with barren materials in order to ensure good mechanical protection, to limit the radiometric impact and to allow a good drainage of surface waters. The sowing of grass on the site facilitates the integration of the disposal and avoids the weathering. The regular supervision with control measurements which occurred during operation are maintained during the reclamation phase and after the phase for several years. For the most part, these measurements concern air and water but also include the local food chain (milk, vegetables, fish, game, etc.)

• Gabon •

URANIUM AND RADIOLOGICAL MONITORING OF THE ENVIRONMENT AT THE COMUF MINING SITE IN MOUNANA

Introduction to the COMUF mining and industrial site

Gabon lies astride the equator on the Atlantic coast of Africa, and has an area of 267 000 km². Structurally the country can be divided into three main areas closely bound up with geology.

Geology

The uranium deposits are situated in south-eastern Gabon in the upper Ogooue province, in the central part of the Franceville basin (see Figure 1).

The central depression where the township of Mounana and the industrial site of the uranium mining company COMUF are located consists of the modified granite-gneiss basement.

The western edge of the Mounana slot consists of Precambrian quartzite sandstone (FA), directly overlying the basement. The Oklo-Okelobondo uranium deposits that are currently being worked lie at the summit of the FA quartzite sandstone near the contact with the pelites FB1 (see lithology chart in Figure 2). They lie on the south-western edge of the Precambrian Franceville sedimentary basin, consisting of a detritic and volcano-sedimentary, non-metamorphic sedimentary series. This series which consists of 5 formations indexed from FA at the bottom to FE at the top lies across an Archean basement.

The sandstone-conglomerate formation FA contains the uranium mineralization, mostly pitchblende, associated with organic matter or occasionally with coffering.

The pelitic FB at the bottom evolves to more sandstone rocks at the top. The other formations, FC, FD and FE are mainly volcanic-sedimentary.

Method for evaluating the environmental impact

In its implementation of a policy for the radiological monitoring of the mining site and the environment, and for evaluating the impact of the uranium ore extraction and processing installations on the population living nearby, COMUF is applying the French regulations (Decree No. 90-222 of 9 March 1990) and International Commission for Radiation Protection, ICRP, Recommendations 26 and 43, until the national Gabon regulations, now under preparation, are ready.

In application of the above regulations, COMUF has set up a system for the radiological monitoring of air and water, the main vectors of radiological pollution in the environment.

As far as the public is concerned, the radiological hazards to be considered are external exposure due to gamma radiation; internal exposure due to the inhalation of short-lived α-emitting daughters of radon-222 and radon-220; internal exposure due to the ingestion of radium-226 and uranium-228 present in local water.

Figure 1. **Geology of the Franceville Bassin**

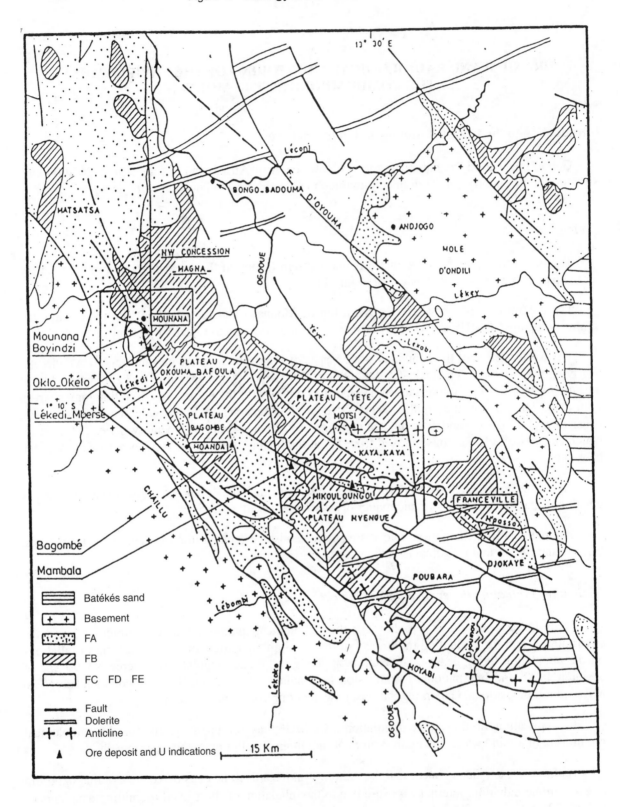

Figure 2. **Lithology of the Franceville Series in the Franceville Basin**

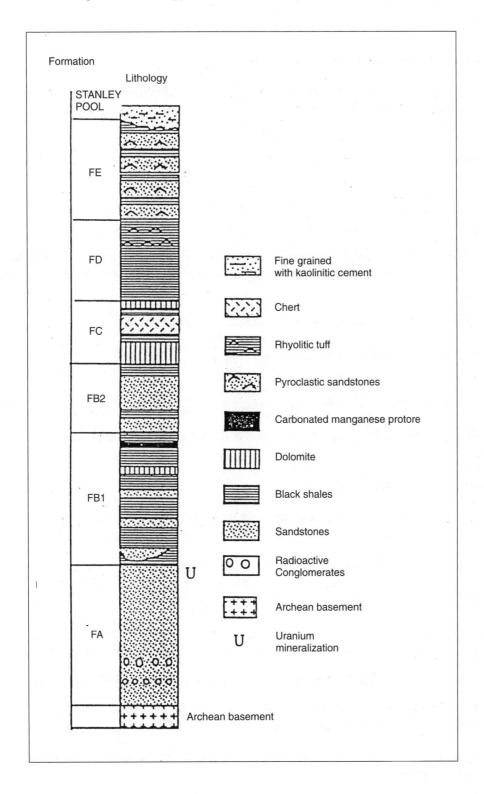

This monitoring system consists of fixed stations and meets the requirements laid down by the regulations cited.

Method for the radiological monitoring of air as a vector

Fixed stations for measuring external and internal exposure due to radon daughters are now in place throughout the zone (see Figure 3). They are used to make monthly measurements of the radiological impact of the industrial facilities on the public.

The monitoring network consists at present of eleven (11) site dosimeters and are distributed as follows:

5 dosimeters located in housing zones close to the mining site:

Massango village; the township known as Ambie or "5 000"; renovation township; Mounana market, replacing the instrument that was installed at the secondary school; and senior staff township.

1 dosimeter located in a housing area remote from the mining site:

Omoi village, regarded as a natural area.

3 dosimeters situated in areas used by workers:

The BTIG factory (sawmill); garage used for cars and trucks; and hospital/offices.

2 dosimeters situated on footpaths, etc.:

Near the old Mounana quarry: an area undergoing renovation through which people pass on their way to offices and the hospital; and below the dike: route to the plantation zones.

External exposure is due to beta and gamma radiation arising from solid substances (heaps of ore and processing plant tailings). The external exposure hazards arise from radionuclides in the uranium series, principally (in decreasing order) ^{214}Bi, ^{214}Pb, ^{234m}Pa and ^{234}Th.

The internal exposure hazards arise from the inhalation of alpha-emitting substances from the uranium 238 series. Two types of radionuclides are likely to be inhaled by the public:

1) Short-lived alpha-emitters: ^{218}Po and ^{214}Po for ^{222}Rn; ^{216}Po, ^{212}Bi and ^{212}Po for ^{220}Rn.

2) Long-lived alpha-emitters present in dust suspended in the atmosphere that may contain traces of uranium and hence of its daughters: ^{238}U, ^{234}U, ^{220}Th, ^{226}Ra and ^{210}Po.

All these criteria are measured by the fixed dosimeters in the monitoring network. Each measuring station comprises:

- A thermoluminescence dosimeter comprising 3 lithium fluoride tablets to measure ambient gamma radiation for 3 months.

- An alpha "measurement head" to measure the concentration in the atmosphere of the potential alpha energy of short-lived ^{220}Rn and ^{222}Rn daughters, together with the total alpha activity of long-lived radioactive dust over a one-month period.

- An air sampler with integral volume metre, which works in conjunction with the alpha measurement head continuously to sample the air that the public might breathe in.

Method used for the radiological monitoring of water

Throughout 1996 and the previous years, COMUF monitored the radiological, physical and chemical qualities of surface water and drinking water using a monitoring scheme that forms part of the programme for checking the impact of its facilities on the environment.

Within the study called SYSMIN, carried out by the consulting company ALGADE in co-operation with COMUF, this monthly monitoring programme was carried out from January to March 1997 using 11 stations covering all the water discharged from the installations into the environment and into different watercourses (see Figure 3).

As part of the radiological monitoring of the COMUF industrial site environment, and for the SYSMIN study, this monitoring of discharged water and the receiving environment was supplemented by quarterly analyses of the drinking water in the different housing areas close to the site over a year (from January 1996 to January 1997).

The analyses performed on all the samples taken covered concentrations of soluble and insoluble ^{226}Ra; concentrations of soluble and insoluble ^{238}U; pH; suspended matter; total salinity; and concentration of sulphates SO_4^{2-}.

The physical and chemical criteria were analysed by the COMUF laboratory using conventional methods.

The radiological criteria were analysed by the ALGADE company. Analysis of ^{226}Ra was done by "emanometry" developed by the French research organisation "Atomic Energy Commission", called CEA/CETEMA method No. 18 of February 1991 and ^{238}U was analysed by "fluorimetry" using the CEA/CETEMA method No. 4 of June 1990.

Results obtained and arrangements for protecting the public

Monitoring air as a vector

Internal exposure due to potential alpha energy (PAE)

At the end of 1996 some ten stations were available for making monthly measurements of the volume concentrations of potential alpha energy in the air due to short-lived radon daughters that could be inhaled by the public.

Table 1 shows all the results recorded at the stations operated by the COMUF radiation protection section since the introduction of environmental radiological monitoring at the Mouana mining site.

Figure 3. **COMUF-MOUNANA**
Scheme for radiological monitoring of the environment
Situation at end 1996

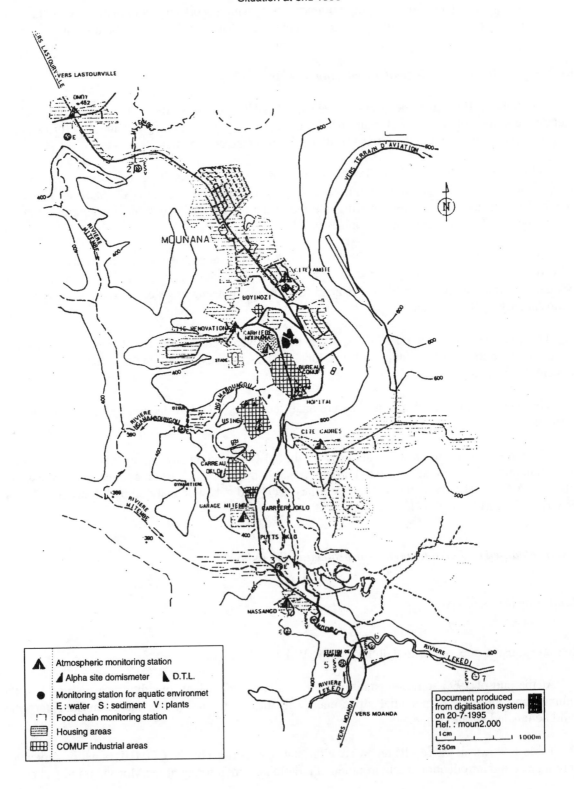

Atmospheric monitoring station
Alpha site domismeter   D.T.L.
Monitoring station for aquatic environmet
E : water   S : sediment   V : plants
Food chain monitoring station
Housing areas
COMUF industrial areas

Document produced
from digitisation system
on 20-7-1995
Ref. : moun2.000

Table 1. **PAE monitoring points for ^{222}Rn (in nJ/m^3)**

YEAR mean value	C 5000	Renovation Township	Garage	Hospital/ BRP	Massango	Omoi*	Mounana Quarry**
1989	96	96		101	54		
1990	110	129		121	72		
1991	23	44		151	33		
1992	145	169	152	125	121		
1993	142	156	284	173	128		
1994	126	183	185	136	100		
1995	97	132	142	114		66	339
1996	87	111	88	111	76	49	
Average	103	127	170	130	75	58	339

* Natural levels.
** Storage area for processing tailings under renovation.

The annual mean level obtained is 121 nJ/m^3 for the PAE for ^{222}Rn. This is below the reference value (285 nJ/m^3) specified in the French Decree No. 90-222 of 9 March 1990 and below the future reference value of 242 nJ/m^3. The overall mean value does not take into account the values from the Omoi site (natural environment) and from the Mounana quarry (storage area for processing tailings, now under renovation).

External exposure due to gamma radiation

By the end of 1996, there were some ten stations equipped with thermoluminescent dosimeters for making 3-monthly measurements of dose rates due to gamma radiation likely to be absorbed by members of the public. The following table gives all the results obtained from the different measurement stations.

Table 2. **Monitoring points for external exposure (gamma exposure monitoring in nGy/h)**

Year mean value	C 5000	Renovation township	Garage	Hospital/ BRP	Massango	Omoi*	Mounana Quarry**	Senior township	Secondary school	Workers township
1989	150	150		180	170			165	160	170
1990	100	100		100	80			100	80	80
1991	140	130		165	115			145	130	140
1992	90	70		120	105			85	90	95
1993	110	125		165	120			140	135	170
1994	120	110	170		120	170		130	180	
1995	160	150	170	150	180	180	870	110		
1996	190	150	180	160	200	200	740	140		
Average	132	123	173	148	136	183	805	127	129	131

* Natural levels.
** Storage area for processing tailings under renovation.

The mean level (137 nGy/h) from all the stations located in the immediate environment of the mining site does not take into account the values obtained at the Omoi site (natural environment) or from the Mounana quarry (storage area for processing tailings, now under renovation).

The results of monitoring water from the Boyindzi, Gamamboungou and Mitembe watersheds, which is not used for domestic purposes, showed the following:

Boyindzi basin

- Absence of soluble ^{238}U in the water of this river. The same applies to ^{226}Ra.

- A concentration of insoluble ^{226}Ra above the detection threshold, i.e. 0.02 Bq/l at 8 sampling points; however these values are still below 0.07 Bq/l.

- The values for pH, suspended matter, sulphates and total salinity are similar to the natural values for the Mitembe watershed.

Gamamboungou and Mitembe basins

Some slight effects of industrial activity can be found in the Gamamboungou and Mitembe rivers below the confluence. The concentrations of soluble ^{226}Ra are between 8 and 15 Bq/l in the Gamamboungou and between 1 and 3 Bq/l in the Mitembe below the confluence. Monitoring showed the sources of soluble ^{226}Ra to be the treatment tailings dike and water extracted from the underground mine workings. A concentration of soluble ^{238}U above 2 mg/l was found at the dike outlet. In the Mitembe watershed, ^{238}U present only in the water extracted from the underground mine workings. Water from the mines is the only source of insoluble ^{238}U.

Water for domestic use

The annual monitoring campaign shows that the drinking water contains no soluble or insoluble radium or uranium. The pH value is close to neutral and the concentrations of sulphates and suspended matter are very low.

Tables 3 and 4 show the results of analyses performed in the underground water and water supplied to villages around the site.

Table 3. **Results from underground water analysis**

SAMPLING STATIONS					
ELEMENTS	EXO 1	EXO 2	EXB	NGA	E5
Soluble ^{226}Ra (Bq/l)	0.52	0.30	0.54	0.05	0.45
Insoluble ^{226}Ra (Bq/l)	38.87	0.02	0.02	<0.02	12.03
Soluble ^{238}U (mg/l)	0.20	<0.10	<0.10	0.10	0.24
Insoluble ^{238}U (mg/l)	2.79	<0.10	<0.10	<0.10	0.78

EXO 1: OKLO extraction drain 400   **EXO 2**: OKLO extraction MCO
EXB: BOYINDZI extraction   **NGA**: Artesian well at Ngangolo quarry
E5: Outlet from OKLO quarry

Table 4. **Results from analysis of water supplied to villages around the site**

SAMPLING STATIONS				
ELEMENTS	Mounana	Omoi	Massango	Ngangolo
Soluble ^{226}Ra (Bq/l)	<0.02	<0.02	<0.02	<0.02
Insoluble ^{226}Ra (Bq/l)	<0.02	<0.02	<0.02	<0.02
Soluble ^{238}U (mg/l)	<0.10	<0.11	<0.10	<0.10
Insoluble ^{238}U (mg/l)	<0.10	<0.10	<0.10	<0.10
PH	7.8	7.2	7.3	6.0

Conclusions

As things stand at present on the Mounana industrial site, water from the Gamamboungou river is unhealthy and unsuitable for any domestic use from the radiological, physical and chemical standpoints.

Water in the Mitembe river below the confluence is acid as far as the confluence with the river Lekedi, and is therefore unsuitable for any domestic purpose.

Water in the Lekedi has no radiological, physical or chemical characteristics that call for special arrangements. It does not appear to present any health risk.

Water being consumed at present does not appear to present any health risk, from the radiological or physical/chemical standpoints to the public concerned living within the site.

The air near the Mounana mining site and its surroundings does not appear to present any particular health risk to the public from a radiological point of view. The results recorded so far appear to be satisfactory and to meet the standards of both ICRP 26 and the French regulations (Decree No. 90-222 of 9 March 1990).

However arrangements are already in hand for renovating the mining site in order to satisfy the future standards (ICRP 60 and 65) with regard to both the public and the environment.

The existing fixed stations on the Mounana site will be retained even after closure of the COMUF installations. This will enable the Mines Directorate, through the regulatory authority, to continue monitoring the radiological situation of this site on completion of the substantial renovation works that are already in hand.

REFERENCES

Radiological Monitoring of Workers and the Environment, Report for 1994 – CRPM-ALGADE Report. RP/95-581/cv. COMU/1.3.

Radiological Monitoring of Workers and the Environment. Report for 1995 – CRPM-ALGADE Report. RP/96-1100/cv. COMU/1.3.

Surveillance and Radiological Protection of Personnel and the Environment at a Mining and Industrial Site. The case of the COMUF uranium mines at Mounana (Gabon). Unpublished report, C. J. LOUEYIT, March 1994.

Storage of Wastes and Renovation of Old Storage Areas at the COMUF Installations at Mounana (Gabon). Report on INITEC expert mission to Gabon (13-19 November 1995).

Study of the Arrangements for Uranium Ore Treatment Tailings Storage and Closure of the COMUF Company's Facilities at Mounana (Gabon). ALGADE, 25 May 1997.

• Germany •

BASIC FACTS

Between 1946 and 1990, 216 000 tU of uranium were produced by the former Soviet-German company SDAG Wismut, third largest producer in the world.

A total volume of 240 million tonnes of ore and 760 million tonnes of waste rock were produced from numerous underground and open pit operations.

The total area affected by uranium mining was 240 km^2 in the Erzgebirge (Saxony) and eastern Thuringia (see Figure 1).

After the unification of Germany in 1990, commercial production of uranium ceased at the end of the year. Wismut still held mining and milling sites in an area of 37 km^2 at the end of 1990 (see Figure 1) and at the beginning of 1991, decommissioning started in areas held by Wismut. The Federal Office of Radiation Protection (BfS) evaluated the extent of uranium mining outside of the area held by Wismut.

On 12 December 1991, the Wismut Act made the Federal Republic of Germany sole shareholder of Wismut, represented by the Federal Ministry of Economics (BMWi).

Decommissioning and rehabilitation

From 1946 to 1990 only minor rehabilitation was carried out. Areas of abandoned mine sites were returned to local authorities but the extent of decommissioning was either limited, not well documented or unknown. Waste rock has been used for the refill of mined-out parts to a certain extent. In 1991 a programme was initiated to evaluate the extent to which clean-up activities will be necessary. Based on Article 3 of the Treaty on German unification, the Federal Office of Radiation Protection (BfS) was commissioned to evaluate all areas in East Germany where uranium mining was conducted (outside the area held by Wismut), whereas Wismut was commissioned to conduct all decommissioning and rehabilitation in its actually mining area (37 km^2) under the supervision of the Federal Ministry of Economics.

Legal framework

The legal framework for the decommissioning and rehabilitation work is defined in a number of laws and regulations, of which the most prominent is the Wismut Act, through which the legislature has charged Wismut GmbH to shut down the former mines and to rehabilitate the associated sites. Other regulations concerning the Wismut-project include:

- The Federal Mining Law.
- The Atomic Energy Act.
- The Federal Emission Protection Law.
- The Environmental Liability Act.
- Legislation for radiation protection in former East Germany, which was carried over by the reunification agreement and is still in effect for uranium mine rehabilitation in Eastern Germany.
- The Radiological Protection Ordinance.

Because there were no significant uranium mining and production activities in the former Federal Republic of Germany, the Radiological Protection Ordinance covers the special situation of uranium mining operations only marginally. Exploration, mining and processing of radioactive mineral resources were included in the ordinance only for completeness and specific requirements and conditions of uranium mining operations were not taken into account. Therefore, the ordinance was not fully applicable to the remediation of waste rock piles, tailings ponds and facility areas. For that reason, the German unification treaty stipulates that the former GDR's regulations on radiation protection should continue to apply to the decommissioning and remediation of uranium mining. In addition, the German Commission on Radiological Protection has issued a number of recommendations on radiological protection principles to be used for the release of contaminated areas, waste rock piles, structures and materials originating from uranium mining. These principles are being considered when assessing the need to remediate areas and facilities. They are described in detail in "Radiological protection principles concerning the safeguard, use or release of contaminated materials, buildings, areas or dumps from uranium mining" in the "Recommendations of the Commission on Radiological Protection" Vol. 23, 1992. These radiological protection principles include recommendations concerning:

- The release for industrial use of areas contaminated by uranium mining.
- The safeguarding and use of mine dumps.
- The release for further commercial or industrial use of buildings used for these purposes and The disposal of building debris from uranium mining and milling.
- The use for forest and agricultural purposes and as public gardens (parks) and residential areas of areas contaminated from uranium mining.
- The release for general use of reusable equipment and installations from uranium mining.
- The release of scrap from the shut-down of uranium mining plants.

These recommendations further include in summary the results of:

- Radiation exposure from mining in Saxony and Thuringia and its evaluation.
- Radiological protection principles for the limitation of the radiation exposure of the public to radon and its daughters.
- Epidemiological studies on the state of the health of inhabitants of the mining regions and the miners in Saxony and Thuringia.

Organisation for decommissioning and rehabilitation of uranium production facilities (modified after BMWi Documentation No. 370)

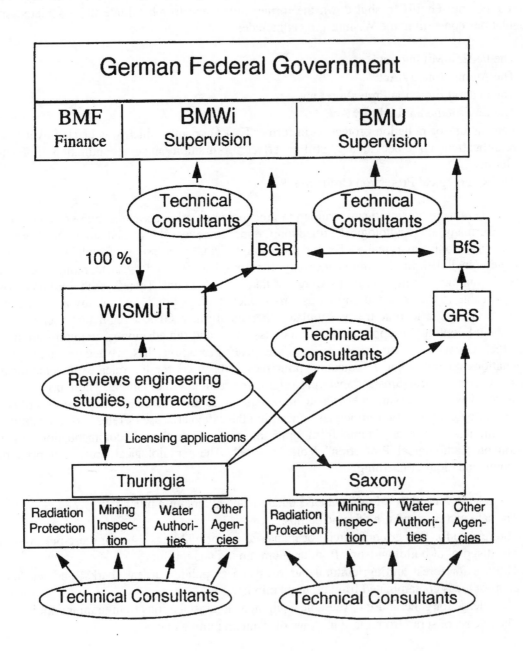

Organisational structure

The following brief overview illustrates the interrelation among the various agencies and corporations engaged in the work for decommissioning and rehabilitation.

- The Federal Ministry of Economics (BMWi) administers Wismut GmbH, a company owned by the federal government, acts as its shareholder and funds the company by annual appropriations from the federal budget after consultation with the Federal Ministry of Finance (BMF) and approval of the German parliament.

- The Federal Ministry of Economics has contracted an expert consultant to assist the project and advise the ministry on proper cost-effective planning and implementation of decommissioning and rehabilitation measures. Use is also made of expert knowledge from Geological Survey, the Bundesanstalt für Geowissenschaften und Rohstoffe (BGR).

- Wismut GmbH is responsible for the decommissioning and remediation of its sites and facilities. Whenever the need arises, the company issues contracts to technical contractors and engineering firms. License applications have to be submitted to the relevant authorities of the State for all decommissioning and remediation operations.

- The licensing authorities in Saxony and Thuringia are in charge of licensing procedures by their respective technical agencies. To the extent required, they also relay on independent expert knowledge.

- The co-ordination of remediation planning with the local communities is of particular importance. The ideas of the local communities regarding the reuse of rehabilitated land are to be taken into consideration within the bounds of economic possibility. Periodic meetings between Wismut management and community representatives serve that purpose.

- The Federal Ministry of the Environment, Nature Conservation and Nuclear Safety (BMU) is responsible for the supervision of licensing actions by the States of Saxony and Thuringia in matters concerning the Radiation Protection Law. To that end, BMU has also contracted a technical consultant for expert services. It also relies on support from the Federal Office for Radiation Protection (BfS).

- The Federal Office for Radiation Protection (BfS) is responsible (under the provisions of the German reunification treaty) for registration of all radioactive wastes in the new States of the reunited Germany including any radioactive contamination outside the Wismut sites. For this purpose BfS started in mid-1991 the project "Radiological registration, investigation and evaluation of contaminated sites due to uranium mining". This project is funded and supervised by the BMU. The project work is conducted by the Company for Reactor Safety (Gesellschaft für Anlagen und Reaktorsicherheit, GRS) and sub-contractors. The results obtained so far are summarised below:

On the basis of recommendations from the German Commission on Radiological Protection, the evaluation was conducted with special emphasis to 1 mSv/year dose rate as the threshold value for individual exposure of the population, additional to the pre-mining exposure. The study areas were defined and field inspections with measurements of radioactivity were carried out. About 5 000 abandoned mining related sites, originating from medieval silver mining and later base metal and uranium mining, were identified in an area of 1 500 km^2. The evaluation reduced this area to about 250 km^2 which will require further investigation and clean-up measures. The comprehensive evaluation was planned to be finished by 1997.

Figure 1. **Mining and milling sites operated by Wismut**

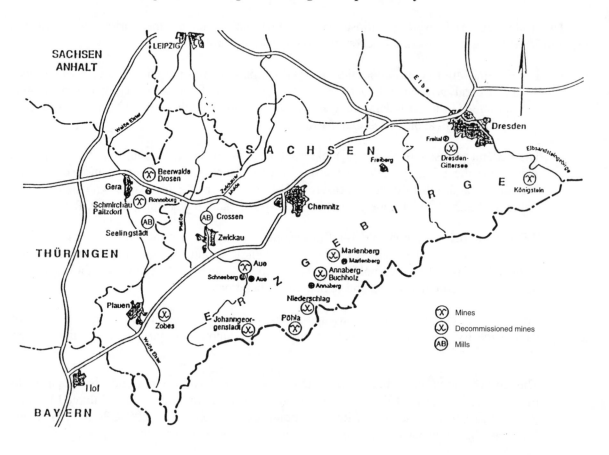

The state of decommissioning and remediation of the Wismut facilities that are closed are described in the Federal Ministry of Economics documentations (BMWi Documentation Nos. 335 and 370).

Major selected rehabilitation issues are summarised below:

- Open pit mining has produced about 600 million m³ of ore and waste rock, of which about 160 million m³ were taken from a single open pit (Lichtenberg), nearly 200 m deep, with an area of about 1.6 km². After mining activities ceased the open pit was used for disposal of waste rock from the Ronneburg mines. About 80 million m³ had been backfilled before 1990. About 100 million m³ of waste material remain piled up, around Ronneburg. Most of this material will be backfilled in the open pit.

- Underground mining has yielded about 300 million m³ of material, half of which was ore. In the Aue district of the Erzgebirge, there are about 40 piles with a total volume of over 45 million m³ covering an area of about 3 km². A major programme of stabilisation, reshaping, covering and revegetation of these piles is being carried out.

- In addition to numerous shafts, drifts totalling about 1 400 km have been driven.

- Mill tailings: two major conventional mills were in operation (Crossen and Seelingstädt). In addition, a number of smaller ones operated for a short period of time at mine sites.

Mining and milling sites of WISMUT

	Mining district			Milling site		Total
	Aue	Königstein	E-Thuringia	Seelingstädt	Crossen	
Area (km^2)	5.7	1.5	16.7	13.1	2.9	37
Shafts	8	10	38	0	0	56
Waste Dumps						
Amount	20	3	16	9	1	49
Size (km$^{2)}$	3.4	0.4	6.0	5.3	0.2	15.3
Volume (10^{6}m^3)	47.2	4.5	187.8	72.0	5.6	317.1
Tailing Dumps						
Amount	1	3	3	7	3	17
Size (km^2)	0.1	0.05	0.9	7.1	2.3	10.45
Volume (10^{6}m^3)	0.75	0.2	0.25	149.3	56.7	207.2
Underground Mines						
Size (km^2)	30.7	7.1	73.4	0	0	111.2
Workings (km)	240	112	1 043	0	0	1 395
Open pits						
Amount	–	–	1	0	0	1
Size (km^2)	–	–	1.6	0	0	1.6
Volume (10^{6}m^3)	–	–	84	0	0	84

- At the Crossen mill, alkaline leaching methods were used for ore from the Erzgebirge. The tailings were disposed of in a nearby pond, about 2 km^2 in size, which now contains about 45 million m^3 of tailings and 6 million m^3 of water.

- The Seelingstädt mill used both alkaline and acid leading methods and mainly processed ores from the Ronneburg district in Thuringia.

- The tailings were disposed of in two nearby ponds with a total volume of 107 million m^3 covering an area of 3.4 km^2.

The total costs for the decommissioning and rehabilitation of all Wismut facilities were estimated at 13 000 million DM and will take about 15 years.

Radiological limits

General

Radiation: 5 mSv/year. Radon (indoor): 250 Bq/m^3. Specific limits for the release of various nuclides are available but are too long to report individually.

Areas of prior use for mining and milling (after reclamation)

Soil (dry)	0.2 Bq/g unlimited use; 1 Bq/g for industrial use only.
Land surface	1 mSv/year or 0.3 µSv/h unlimited.
Debris of buildings	0.2 Bq/g unlimited; > 1 Bq/g restricted use.
Scrap of uranium production	0.5 Bq/cm^2 unlimited.
Mine dumps	0.2 Bq/g unlimited or < Bq/g at < 1 ha or < 105 m^3 unlimited.
Recommended limit for mine workers or other personnel exposed at work places	50 mSv/year (100 mSv/5 years).

• Hungary •

CONCEPTUAL PLAN FOR CLOSURE OF HUNGARIAN URANIUM INDUSTRY IN THE MECSEK

The Hungarian Government decided to close its uranium industry on 31 December 1997. During 1996, both Mecsekuran Ltd. and the Mecsek Ore Mining Co. prepared the Conceptual Plan for the closure of the uranium industry in the Mecsek Mountains. This plan was reviewed and accepted by the competent authorities (mining, environmental, water, etc.). It considers all the facilities in the 65 km^2 area including 4 transportation shafts (110-1 200 m), 3 ventilation shafts, the main transportation gallery to the surface and the open underground workings of the mines.

The Conceptual Plan specifies procedures for dismantling and demolition of the ore processing plant, including emptying and decontaminating the technological structure. The Plan defines the methodology for the remediation works: soil and water investigations, country planning, drainage, covering the tailing piles and ponds, indicates the thickness of the new soil and vegetation, etc. In addition, it provides a timetable, together with a very detailed financial plan for the work. A very important part of the Plan is the "human-question".

Some of the issues contained in the Conceptual Plan are:

- Regulatory requirements (authorities, basic principles and radiological requirements).

- Inventories of the open underground mine workings, buildings and facilities, tailings ponds and heaps, heap-leaching heaps and remaining uranium reserves and resources.

- Description of the deposit and mining and milling facilities (the geological and hydrogeological environment and the ore processing).

- Impacts of the uranium mining and milling (problems are classified according to urgency).

- Concepts for the Remediation (water treatment, uranium treatment and monitoring).

- Concepts for re-utilisation of the mining and milling areas.

In April 1997, the Hungarian Government decided to accept the closure of mines according to this Plan. The closing of works, remediation and country planning must be finished by 2002 and the expected cost is HUF 14 to 15 billion, equivalent to about USD 80 to 85 million.

The relevant regulations are: Mining Law No. XLVIII/1993; Governmental Regulation No. 152/1995 about the environment; Act on Atomic Energy 1996; IAEA Technical Report No. 362; and 1991 Report of the National Radiobiological and Health Institute.

• India •

Environmental issues in uranium mining and milling in India are controlled by the 1996 Atomic Energy (Factories) rules, and the 1996 Atomic Energy Radiation Control rules, within the framework of the 1948 Atomic Energy Act, further amended in 1962. A safety committee set up by the Atomic Energy Regulatory Board oversees the safety aspects of the projects of Uranium Corporation of India Ltd. (UCIL) and the Atomic Minerals Division (AMD).

Mines

Adequate ventilation is maintained in the mines to keep radon and dust particles within permissible limits. Run of the mine ore contains 60-64% silica and has a uranium content of 0.05 to 0.065% U_3O_8. Exposure to airborne silica dust is a perceived hazard. Radiological hazards from exposure to airborne long-lived alpha activity, radon and external gamma radiation, although present, are of lesser hazard due to the low grade of the ore.

Dust control is effected by adopting Dry Fog System and Wet Scrubber connections in the crushing section. All personnel have been provided with protective equipment such as ear muff respirator, safety helmets, leather shoes, gum boots and protective clothing.

Mills

Uranium ore lumps are reduced in two stage crushing from -200 mm to -25 mm. The final product from the Triple Deck Screen goes for grinding with some -113 to +65 mm pieces for use as autogenous grinding media.

Crushed ore is reduced from 608 to 200 mesh in two stages by wet grinding. In the grinding circuit a dust suppression system is provided at the fine ore transfer points. Cleaning of ore spillage and slurry is carried out on a regular basis in the working area. Ore slurry is fed to pachucas for acid leaching with sulphuric acid and pyrolusite at 40°C. No manual handling of sulphuric acid is involved. Pyrolusite is wet ground and handled as a slurry. Uranium is recovered from the clarified leach liquor by Ion Exchange and precipitated as ammonium diuranate.

The neutralised tailing slurry is classified in hydrocyclones. Coarse sand is sent to the mines for backfilling. Fines are pumped to the tailings pond for containment. Excess liquor is decanted. A portion of it is recycled to the plant and the rest is treated with barium chloride for precipitation as radium barium sulphate and with lime at pH 10 to precipitate manganese. Settled precipitates are pumped back to the tailings pond. The pH of the treated water is adjusted to 7.5-8 with sulphuric acid and released to the natural stream.

• Japan •

Currently no uranium mining and milling is carried out in Japan. Wastes from past uranium production activities have been controlled in accordance with all related regulations.

Regulations applicable for the shut-down of uranium mines and mills

- Uranium mining facility: Mines Safety Act and Mine Safety Regulations.

- Uranium milling facility: Mines Safety Act and Mine Safety Regulations.

- Nuclear raw materials: Nuclear Fuel and Nuclear Reactor Act (Atomic Control Act) and regulation for nuclear raw materials and nuclear fuel materials milling operation.

Mine safety act and regulations

The regulations are not only aimed at uranium mining and milling but also at commodities such as oil, coal, limestone and metal mines. The regulations include radioactive safety clauses for uranium mining and milling to prevent radiation hazards.

If the mining right is transferred or abandoned, the "Mine Safety Act" requires that responsibility is established to take measures to prevent mine pollution from a waste rock dump, tailings disposal site or an abandoned excavation. The act also establishes the entity which is responsible for this issue.

A holder of mining rights is only allowed to dispose of waste rock, tailings, and precipitates using a form of piling and layering. Technical standards for reclaiming wastes are established in the regulation.

Waste rock and tailings are allowed to be disposed of as piles. Protective walls and dams to prevent collapse of waste effluents are required. After piling of the wastes is terminated, a disposal site must be covered by soil and revegetated.

The above regulations are applied to all mines, including uranium mines. In a uranium mine the level of external radiation and radioactive materials in the air and water are of special concern. They are regulated to prevent radiation hazard to both workers and the public. Nothing is mentioned following closure of a facility. It is interpreted, however, that the regulations for preventing radiation hazards to humans for an operating facility apply to the closing of a facility.

This *Atomic Control Act* is aimed at (1) restricting uses of nuclear raw materials, nuclear fuel materials and nuclear reactors to civil purpose only, (2) preventing disasters and (3) securing public safety. This act applies only to milling, not to exploration and mining activities.

When a milling facility is closed, the operator removes all nuclear fuel material and decontaminates the facility, and disposes the contaminated materials. The regulatory minister requires that necessary measures be taken if the close down actions have not been carried out properly. When contaminated nuclear fuel materials are disposed outside the facility, safety measures are required under the ministerial ordinance. The minister will require necessary measures to be taken when the manner of disposal violates the ministerial ordinance.

• Jordan •

Jordan has four phosphate deposits containing uranium. At present, there is no production of uranium. However, a feasibility study for extraction of uranium from phosphoric acid has been introduced.

To assess the environmental effects of uranium bearing phosphate deposits, a systematic study and evaluation of the uranium concentration in Jordanian phosphates is being conducted. The Shidia phosphate deposits, which constitute by far the largest phosphate reserves in Jordan, are characterised by relatively low levels of contained uranium, i.e. average 50 ppm. This is judged to be highly positive from the environmental point of view regarding raw phosphates, derived phosphoric acid and phosphatic compounds, and derived fertiliser products.

• Kazakhstan •

Over the last 50 years more than 100 uranium occurrences have been explored and about 100 of them have been mined. In association with this activity, about 230 million tonnes of radioactive waste have been accumulated in Kazakhstan. These wastes are different in activity and aggregate state.

Kurday was the first economic uranium deposit discovered in Kazakhstan. The discovery of the deposit in southern Kazakhstan in 1954 was soon followed by mining. For processing ores from this deposit, and other ores from southern Kazakhstan, the Kirgiz Production Centre (mine-ore combinate) was put into operation in 1959. Now the Kirgiz Production Centre, together with the process tailings are situated in the territory of the Republic of Kyrgyzstan. In later years, three production centres were established in different parts of Kazakhstan: Pricaspian Centre in the west, Tselinny in the north and Ulbinsky in the east.

In Kazakhstan there are about 15 worked out, shut down or conserved uranium deposits. Currently, uranium production is considerably reduced and uranium is mainly being produced by the ISL method. This method creates less radioactive waste than conventional mining. However, Kazakhstan still has a problem of abandoned radioactive waste, because the companies that produced these wastes no longer exist.

Currently, only two production centres (Tselinny and Ulbinsky) are in operation and still maintaining their tailings. The Pricaspian Production Centre is closed. Its tailings have not been maintained. They present a danger, such as "dusting beaches" for example.

In Kazakhstan there is a law on the "Use of Atomic Energy" and a draft law on "Radwaste Management". There are regulatory codes on decommissioning and rehabilitating production areas. There are several projects for rehabilitating areas with mining and milling waste. However, all of these projects are not being carried out due to financial difficulties in Kazakhstan.

The systematic investigation and inspection of mine waste started in Kazakhstan in the 1990s. In 1996, the first stage of the "Assessment of Urgent Measures to be taken for Remediation at Uranium Mining and Milling in the Commonwealth of Independent States, Regional Project No. G42/93, Project No. NUCREG 9308" was carried out in the framework of the European Union support programme to former CIS countries, TACIS. The cataloguing of 100 waste storage sites has been completed. Five sites were chosen to carry out a survey including measuring the influence of the waste on the nearest settlements. During completion of the 3rd and 4th stages of this project the modelling and development of measures for rehabilitation of the whole territory of Kazakhstan will be formulated.

Since 1995 uranium has mainly been produced in Kazakhstan using ISL technology. Considerably less radwaste is produced, but other problems are associated with the method. One of the problems is contamination of ore-bearing aquifers in association with the uranium extraction. There are different viewpoints on possible restoration of the aquifer. There is the opinion that the aquifer could undergo self-treatment or "self-restoration" during 10-20 years following extraction. However, most investigators believe that the rehabilitation of aquifers after extraction is an expensive and complex problem. This is thought to be especially the case in Kazakhstan where sulphuric acid ISL technology is used.

Currently, for the choice of optimal ISL technology in Kazakhstan, the Technical Co-operation Project KAZ/3/002: "Modern Technologies for In-Situ Leach Uranium Mining" is being carried out by the IAEA in co-operation with Kazakhstani organisations. The objective is to decrease the environmental impact of ISL uranium extraction. Results of the project which began in 1997 will be useful for licensing uranium production facilities with the least contamination of ore-bearing aquifers.

• Namibia •

RÖSSING URANIUM MINE

The only operating uranium mine in Namibia is the Rössing Uranium Mine. The information provided below relates to this mine and is extracted from "The Rössing Fact Book".

The Rössing Mine follows a strict code for the protection against radiation. The code is based on the recommendations from the International Commission for Radiological Protection (ICRP) and the International Atomic Energy Agency (IAEA). The primary objective of establishing this code is to ensure that exposures to radiation will not give rise to unacceptable levels of risk. The sources of radiation are identified, quantified, controlled and minimised.

The ore at Rössing is of low grade. The highest concentration of uranium is occurring in the final product, where uranium oxide is roasted and packed. Employees working in this area wear special protective clothing, as well as a personal radiation thermo luminescent dosimeter. Once a month a urine sample from each employee is analysed for traces of uranium as a check for internal exposure. All employees leaving the final product recovery section are monitored with a radiation scanner. The average production worker at Rössing is exposed to about 4.5 mSv per year which is below the 20 mSv limit recommended by the ICRP.

Health and environmental audit

A technical team from the IAEA, the International Labour Organisation and World Health Organisation conducted a comprehensive health and environmental audit at the invitation of the Namibian Government. This report is available from the IAEA and concluded that "Rössing's radiation safety and medical surveillance programmes could serve as a good example to similar industries in the rest of the world".

Radon and thoron concentrations are regularly measured in the mine and surrounding areas. Various studies have shown no abnormal high radon concentrations. Furthermore the thoron concentrations are one eighth that of radon. Sulphur dioxide, welding gas, ammonia, exhaust fumes and the radioactive gases radon and thoron are the common industrial hazards. Sulphur dioxide is the biggest pollutant emitted from the acid plant and several sulphur dioxide detectors are located in the area. The United States Environmental Protection Agency's computer models are used to simulate the emissions. The average emissions are within the South African air quality standards, which are used as guide lines. It is mandatory for all employees coming into contact with chemicals, gas and fumes to wear appropriate protective clothing.

The mine is situated in the dry environment of the Namib Desert and natural dust storms are common. The dust generated at the mine is suppressed at various points throughout the mine.

Blasted rock and the haulways are sprayed with water. Only wet drilling is allowed and the ore is sprayed at the primary crusher. At various stages during crushing the dust is collected mechanically. The cabs of all the mobile equipment are fitted with air conditioners and dust filters. Personnel working in dusty outside areas wear respirators or "Airstream" helmets. The tailings impoundment is the largest single source of dust emissions and therefore an erosion resistant surface is maintained over the impoundment.

Water supply

Water is essential to the mine's mining and milling process and plays a major part in the extraction process. It is also used as carrier of ore and waste. The water is supplied by the Ministry of Agriculture, Water and Rural Development from underground sources. Water management focuses on economising and recycling, as well as the control of seepage water from the tailings impoundment to prevent contamination of the natural ground water. Some 60% of water used by the mine is recycled. The acid and radioactive material is captured by the tailings. Seepage water is stopped by a series of trenches and wells and is pumped back for reuse. Water samples are continuously collected and the quality is monitored in conjunction with the Department of Water Affairs. A vegetation monitoring programme is also conducted.

Decommissioning plan

A decommissioning plan dealing with radioactivity, water and air quality and the rehabilitation of the mine site has been compiled. The mine life is expected to remain in operation for at least another 20 years and the implementation of the decommissioning is thus not immediate. The decommissioning plan is reviewed every 5 years and the budgeted amount is reviewed annually to make provision for inflation. The main elements of the plan are: stabilising the tailings and covering them with a layer of crushed rock; cleaning, dismantling and removing all manmade structures; landscaping the site of the processing plant area and covering it with waste rock and alluvium; and fencing the mine area.

Seepage of water from the tailings dam will continue to be intercepted by the installed control facilities after mine closure. This seepage will be pumped into the open pit where it will be allowed to evaporate.

• Niger •

URANIUM AND THE ENVIRONMENT IN NIGER: RADIOLOGICAL MONITORING AND THE URANIUM ORE MINING AND PROCESSING ENVIRONMENT

Introduction

Uranium mining began in Niger in 1968. Since then about 68 917 tonnes of uranium have been extracted. At present two companies – Somair and Cominak – are mining uranium in the country, at a rate of about 2 960 tonnes a year. This made Niger the world's third biggest uranium producer in 1996, with an output of about 3 320 tonnes. In 1996, 2 077 persons were employed in the uranium sector. These workers, and others involved in related activities, live with their families in two mining towns located less than 10 km from the mines and the uranium processing plants. According to the 1997 census, the total estimated population of these two towns was 46 058.

Owing to the radiation risks to which the workers in the uranium mines and the surrounding population – particularly in these mining towns – are exposed, the two companies have gradually developed radiological monitoring programmes to control these hazards.

To this end each company has set up a radiation protection service (including a radiation protection laboratory), in charge of carrying out radiological monitoring and taking action to ensure that radiation doses to workers and their environment comply with current legislation (Order 31/MM/H of 5 December 1979).

Radiological exposure

Workers in the uranium mines and ore processing plants, and the public living nearby, are exposed to specific radiation risks arising from the uranium and the accompanying radioactive materials. These effects are:

- External exposure due to gamma radiation (γ).

- Internal exposure due to the potential alpha energy (PAE) of short-lived radon-222 daughters and to uranate and ore dust present in the atmosphere.

- Internal exposure due to the ingestion of radium 226 and uranium 238 present in the water and food consumed.

The regulations provide maximum dose limits (see Table 1) for of each of these forms of exposure.

According to Order 31/MM/H, the cumulative total of these 4 risks is calculated using the following formulae:

For workers

$$Cumulative\ total = \frac{\gamma}{150\,mSv} + \frac{PAE\ ^{222}Rn}{14.4\,mJ.m^{-3}} + \frac{Oredust}{2\,590\,Bq.m^{-3}.h} + \frac{Uranate\ dust}{10\,600\,Bq.m^{-3}.h}$$

For surrounding population

$$Cumulative\ total = \frac{\gamma}{5\,mSv} + \frac{PAE\ ^{222}Rn}{2\,mJ} + \frac{PAE\ ^{220}Rn}{6\,mJ} + \frac{Ore\ dust}{170\,Bq} + \frac{^{226}Ra}{7\,000\,Bq} + \frac{^{238}U}{2\,g}$$

Each of these cumulative totals should be less than 1.

Table 1. **Annual regulatory limits (31/MM/H)**

Type of exposure	Regulatory limits	
	Company employees	Surrounding population
Gamma radiation (external): γ	50 mSv	5 mSv
Potential alpha energy from ^{222}Rn (internal): PAE ^{222}Rn	14.4 mJ.m^{-3}.h	2 mJ
Long-lived alpha emitters present in ore dust (internal)	2 590 Bq.m^{-3}.h	170 Bq
Long-lived alpha emitters present in uranate dust (internal)	5 200 Bq.m^{-3}.h	–
Short-lived radon daughters: PAE ^{220}Rn	–	6 mJ
Ingestion of radium 226 present in water and foodstuffs consumed (internal): PAE ^{226}Ra	–	7 000 Bq
Ingestion of ^{238}U present in water and foods consumed (internal): ^{238}U	–	2 g

Stages of radiological protection

Radiological protection can be divided into three stages: prevention, radiological monitoring and treatment of serious cases.

Radiological prevention

The objective is to identify the sources of irradiation so that they can be eliminated or if possible, their effects on personnel and the environment reduced to the strict minimum. The following actions are taken:

- Controlling the spread of radioactive dust: spraying tracks, areas affected by mining blasts, crushing plants, and so on.

- Removing radioactive gases (radon and its daughters) by enhancing ventilation (primary and secondary ventilation, pressure barriers).

- Isolating drum filling compartments, cleaning up areas of high exposure (crushing, chemical processing, filtration).

- Ensuring that workers wear masks to protect themselves against dust.

- Reinforcing the health physics teams.

- Providing exposed workers with foodstuffs that help the body to deal with dust, particularly milk.

- Bodily hygiene and cleanliness of working clothes.

Radiological monitoring

This involves measuring (at the monitoring stations) the radiation doses received by workers, the public and their environment, with a view to reducing the hazards and finding timely methods of minimising the risks of irradiation in accordance with the current regulations. Radiological monitoring involves: personnel dosimetry, radiological monitoring of working areas and radiological monitoring of environmental pollution.

Personnel dosimetry

Radiological monitoring of personnel is based upon the use of the integrated dosimetry system produced by the French consulting company ALGADE and involves the following three methods:

- Individual dosimetry: this consists in providing certain individual employees (the most exposed) with instruments (dosimeters) to measure the radiation dose received. These are underground workers in the mining, geology, management and maintenance groups.

- Functional dosimetry: this consists in monitoring the radiation dose received by an individual taken at random from a group of employees working in the same conditions; each of the particular worker's colleagues is then assumed to have received the measured dose.

- Environmental dosimetry: this consists of monitoring the workplace; each worker in the particular area is then assumed to have received the measured radiation dose in proportion to the time he has spent there.

Table 2. **Personnel dosimetry**

Monitoring method	Number of instruments used		Number of employees monitored		Workers monitored	
	Somair	Cominak	Somair	Cominak	Somair	Cominak
Individual	85	456	–	456	All workers	Underground and plant (the most exposed)
Functional dosimetry Environmental dosimetry	–	69	–	435	–	Temporary in the mine and the plant (the least exposed)
Total	**85**	**525**	**614**	**891**		

The personnel dosimetry system (Table 2) makes it possible to determine the doses received:

- Quarterly for external exposure due to gamma radiation.

- Monthly for internal exposure due to the potential alpha energy of short-lived radon-222 daughters, long-lived alpha emitters present in ore dust, and long-lived alpha emitters present in uranate dust.

The exposure limits authorised by legislation in Niger (Order 31/MM/H) are shown in Table 1.

Monitoring the workplace environment

This is done as follows:

- *Daily monitoring of the working atmosphere.* Spot measurements of the working atmosphere in the mine and the plant are made daily as part of the monitoring of the working environment for at least one of the four hazards, so that immediate steps can be taken wherever there are abnormal risks.

- *Dust samples from settlement plates.* Settlement plates are installed at various points in the working area. The plates are removed and replaced weekly and their contents weighed and analysed.

Radiological monitoring of the environment

This involves units for collecting dust on settlement plates, and analysis of vegetables and water consumed by the surrounding population, and of earth samples. Radiological monitoring of the environment is used to determine any damage caused to the surrounding population so that the necessary steps may be taken to ensure that the limiting radiation doses authorised by current legislation are not exceeded (see Table 1).

The risks incurred by the public are external exposure due to gamma radiation and internal exposure due to inhaling short-lived α-emitting daughters of radon 222 and radon 220; inhaling long-lived α-emitters present in ore dust and suspended in the air; and ingestion of radium 226 and uranium 238 present in foodstuffs consumed.

Accordingly each of the two companies utilises four stations for the radiological monitoring of the environment. The exposed population is sub-divided into two categories:

- Group 1 (G1): people living in the mining towns (ZU).

- Group 2 (G2): people living close to the installations, between the mining township and the industrial area for each of the companies.

Table 3 shows all the points at which the two companies carry out radiological monitoring of the environment. In addition to this, the companies sample and analyse soil along radials passing through their operational zones.

Figure 1.
SOMAIR (ARLIT – NIGER)
Changes in specific radium-226 activity

In vegetables

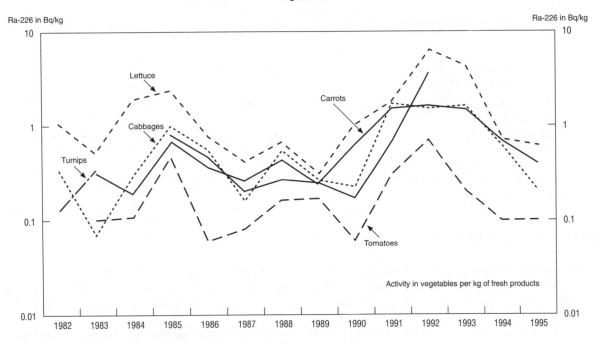

Figure 2.
COMINAK (AKOUTA – NIGER)
Changes in specific uranium-238 activity

In vegetables and earth from the Akokan garden

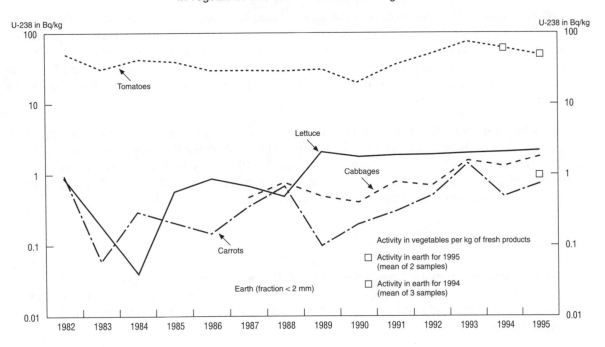

121

Table 3. **Stations for radiological monitoring in the working zones**

Monitoring station	Population type monitored	Location	
		Somair	Cominak
ZI	–	Somair plant	Akouta
ZU	G1	Arlit	Akokan
Two stations located between ZI and ZU	G2	Shaft 214 and REG	Akouta-East and base camp
Sampling of vegetables and earth	G1 and G2	Amidar	Akokan garden
Water sampling	G1 and G2	Arlit water supply	Akokan water supply

Figures 1 and 2 show how measured levels for activity of ^{226}Ra and ^{238}U changed in the period 1982-1995, respectively.

Treatment of serious cases

The regulations require companies to provide half-yearly and annual reports on the radiological situation of workers and the environment. The radiological situation of the environment has so far not been alarming, but there have been a few cases where worker exposure limits have been exceeded.

These cases are dealt with by moving employees who have reached their dose limit to a different job; intensifying preventive measures to reduce exposure at the most affected posts; and reducing the time spent by workers in the high-risk zones.

Government action

To ensure that the legislation on radiological hazards is applied, the Ministry of Mines and Energy has set up a mining section at Arlit (the main mining township) with a radiation protection laboratory and staff who are aware of radiological hazards. This section makes periodic visits to the plants and mines of both companies, takes its own samples and monitors the environment in order to collect its own radiation data; it also issues advice to the employers to ensure that radiological protection is as good as it can be.

In order to obtain information on the effect of uranium mines on the environment, the Arlit mining section will shortly have its own stations (two in number) for the radiological monitoring of the environment. One station will be installed at a point affected by both companies to indicate the total radiological pollution generated; a second station will be installed fairly distant from the working zones of the two companies to obtain data on the natural levels of radioactivity.

Conclusion

Since they were set up, the companies operating the uranium mines in Niger have made and are continuing to make considerable efforts for the radiological protection of their workers. Through the assistance of the IAEA, the mine management has acquired radiation protection equipment and has improved relevant skills.

A National Radiation Protection Centre will shortly be set up in Niger with the help of the IAEA. Its head office will be in Niamey. This Centre will help improve radiological monitoring, particularly by promoting the rapid use of dosimeters and other radiation measuring instruments, as well as the monitoring of all radioactive sources in Niger.

• Portugal •

Portugal has been producing uranium since 1951, mainly at the Urgeirica plant. Empresa Nacional de Uranio S.A., (ENU), has been monitoring several environmental parameters such as air quality, mining effluents (underground mine and surface drainage waters) and collecting data samples on soil, sediments and vegetation for further analysis on the following decommissioning mines: Urgeirica, Castelejo and Cunha Baixa.

Every mine has wells and piezometers. The analysis of underground and surface drainage waters is done at several monitoring sites upstream and downstream every watercourse near the mines.

Underground waters are being monitored at a distance of 300-400 metres beyond the tailings dam perimeter of the Urgeirica plant and the cleared waters are monitored downstream at a distance of 3 km.

On the radiological protection field, several analyses are being carried out to detect any radio-elements in water.

Several studies are being carried out to characterise geochemical and hydrochemical aspects and to establish what effects the waste piles of the Cunha Baixa mine (decommissioned mine) and Quinta do Bispo mine (heap leaching mine) may have had on the environment. Work is also done to establish what mitigation measures, if any, may be required to counter any negative impacts. Thus, alluvial and sediment samples are being collected upstream and downstream on every water course near the site.

• Romania •

Romania has been producing uranium since 1952. The production is presently carried out at three mines and one mill.

The Romanian uranium industry has a systematic programme for protection of the environment. Potential sources of environmental impact during uranium exploration, exploitation and milling activities include:

- Mine and mill effluents containing radioactive elements above the maximum admissible concentration.
- Waste rock from mining operations.

- Low grade ore with a uranium content of 0.02-0.05%, which at present is not processed, but is stored at the mine site.

- Tailings from processing activities, stored in the dewatering ponds at the Feldiora mill.

- Metal and wooden wastes contaminated with radioactivity during exploitation and processing of radioactive minerals.

The installations and equipment required for the prevention of environmental contamination include construction and enlargement of the water treatment plants for treating effluents at the Carpates Orientales, Montagnes Apuseni and Banat deposits; increased capacity of tailings impoundments; processing and closing of the ore storage areas at various mines; long-term stabilisation, reclamation and revegetation of waste dumps and surrounding areas.

The following general activities are planned: rehabilitation of the entire rail car park; implementation of a ventilation system using filters at the mill and ore shipping stations; and equipping all uranium production sites and facilities with environmental monitoring systems.

• Russian Federation •

Two uranium production centres have operated in the Russian Federation: Priargunsky Mining-Chemical Production Company (PPGHO) and Lermontovsky State Company "Almaz".

Priargunskoye Mining-Chemical Production Company (PPGHO)

Working environment

PPGHO is located in the Chita region of the Russian Federation, 10-20 kms from the town of Krasnokamensk. Krasnokamensk has a population of about 60 000. It is the operator of the Streltsovsky uranium ore region, including 19 uranium deposits with an average grade of about 0.2% U, in an area of 150 km^2. Mining has taken place since 1974 in 2 open pits (both are depleted) and six underground mines (3 are active and 3 have been conserved). Milling is carried out at a hydrometallurgical plant using sulphuric acid leaching with subsequent recovery of uranium by ion exchange. Since the 1990s low grade ores are being processed using heap leaching. The staff of PPGHO includes about 10 000 employees. The annual uranium production is about 2 500 tonnes.

Emission into air and water

Monitoring of the hydrosphere and atmosphere is conducted by Priargunsky's laboratory. The main sources of environmental contamination are the tailings ponds of the hydrometallurgical and the sulphuric acid plants. The tailings have a total volume of 300 million cubic meters and include 9 000 Ci of radioactivity.

Characteristics of the wastes

Type of waste	Ha^2	Amount 10^9 tonnes	U cont. %	Radioactivity $x10^{-9}$ Ci/kg	Radon emanation 10^{-3} Ci/m^2/year
Milling tails	600	40.0	0.009	30-750	0.93-23.2
Acid plant wastes	320	5.6	traces	30-250	–
Radiometric sorting wastes*	48	8.0	0.012	27-350	0.84-11.0
Piles of low grade ores*	270	28.0	0.009	27-350	0.84-11.0
Mining waste rock dumps	80	7.0	0.002	27-80	0.84-2.50
Open pit overburden	275	190.0	0.001	27-80	0.84-2.50

* Wastes suitable for heap leaching.

The contents of radionuclides (Ci/l) in ground water around the tailings pond and in mine water vary within wide limits.

Type of water	U (mg/l)	$^{226}Ra \times 10^{-1}$	$^{230}Th \times 10^{-1}$	$^{210}Po \times 10^{-11}$	$^{210}Pb \times 10^{-11}$
Ground water	0-8.7 x 10^{-4}	0-4.3	0.12-5.5	0.03-1.9	0.16-2.4
Mine water*	0.17-21	1.0-39.0	3.0-28.0	1.3-69.0	1.7-58.0

* Average and maximal mean.

Waste management

Currently two main environmental problems exist: increasing accumulations of radioactive liquid and solid wastes; and progressive contamination by radioactive wastes of natural surface and groundwater systems, thereby creating a potential threat to the potable water supply.

The first stage of a mine water restoration plant was built in 1996, based on zeolite sorption technology.

There is also a potential threat of the failure of the mill tailings pond dam, which would result in flooding of Urulungui valley and seepage of wastes to the Urulungui and Argun Rivers.

Completion of the project for reconstructing the waste ponds for both mill tailing and acid plant waste is planned for 1998. It includes:

- Strengthening of main dams and building of a protective dam around the potable water boreholes.

- Installing a duplicate potable water supply source, with a 1 000 m^3/hour capacity, in the Dosatui settlement.

- Constructing a drainage curtain of intercepting boreholes below the tailings pond dam.

Decommissioning and restoration

The shut-down procedure of uranium mines and mills is currently regulated by official state instructions. The Tulukuevsky and Krasny Kamen open pits were the first two depleted deposits of Streltsovsk region. It is suggested that an international programme for their rehabilitation be created.

Lermontovskoye state company "Almaz"

Work environment

The "Almaz" site of the Lermontovskoye state company is located 1.5 km from the town of Lermontov, in the Stavropolsky region of the Russian Federation. The region has two depleted uranium deposits: Beshtau and Byk, with a total production of about 5.3 thousand tU (produced from ores with an average grade of 0.1%). Operation of the two underground mines started in 1950. The Beshtau mine closed in 1975 and Byk closed in 1990. From 1965 to 1989 ore was processed at the Lermontovsky mill using sulphuric acid leaching, as well as by heap leaching. In situ leach mining was also carried out at the site. In addition, from the 1980s to 1991, ore from the Vatutinskoye uranium deposit in Ukraine and from the Melovoye deposit in Kazakhstan was also processed at the Lermontovsky plant. Since 1991, after uranium production was stopped, apatite concentrate is being processed at the plant. In 1996 a radiometric survey of a 3 200 hectare area was conducted.

Emission to air and water

The monitoring of surface, ground and mining water, of residual milling liquids and of the atmosphere is conducted by the company's laboratory. The area of radioactive contamination of the facility includes more than 170 hectares, of which 118 hectares are in the mill tailings pond, and 54 hectares of the milling plant and waste rock dump from mining.

The following table provides information on the waste rock.

Source of wastes	Area of waste (hectares)	Amount of waste (million tonnes)	Contents (% U)	Radioactivity (10^{-9} Curie/kg)	Radon emanation (10^{-3} Curie/m^2/year)
Mining waste rocks	54	8.4	0.002	25-80	0.90-2.50

The main source of environmental contamination is the exhalation of radon-222 from the tailings pond. The contamination of ground water resulted from drainage of fluids from the tailings pond, particularly during the first years of the plants operation. The average and maximal uranium content in the first ground water aquifer below the tailings pond has been controlled using 10 hydrogeological boreholes located around the perimeter of the tailings pond. The results of the monitoring programme are shown in the following table.

U Concentration (10^{-6} g/l)

	1983	1984	1986	1987	1988	1989	1990
Average	40	54	28	94	75	35	39
Maximum	67	76	50	170	90	80	40

The emission of radionuclides to the atmosphere and surface waters in Ci/year, after uranium mining and milling was stopped is summarised:

	Year	Wastes, m³ x 1 000	Total U nuclides	Th nuclides	^{226}Ra	^{210}Po	Total Ra nuclides	Long-lived nuclides
Podkumok River	1990	5 328	0.056	0.081	0.021	0.006		
	1991	1 012	0.115	0.0064	0.005	0.004		
Atmosphere	1990					0.0031		0.0033
	1991					0.00015	0.00003	

Waste management and rehabilitation

Rehabilitation activities are funded from the state budget and are conducted according to state regulations. Rehabilitation of the waste rock dumps of Mine 1 (Beshtau deposit), with a surface area of 36 hectares, is nearly completed. Rehabilitation of the waste rock dumps of Mine 2 (Byk deposit), with a surface area of 18 hectares is underway.

The project of remediating the milling complex (buildings and territories) is in the planning stage. Rehabilitation and decommissioning of mill tailings pond, with a surface area of 118 hectares was started in 1996 and is expected to continue until 2005.

• Slovenia •

Ore production in the Zirovski vrh uranium mine started in 1982. The mill began operation in 1984. In 1990, the production was terminated and placed on temporary stand-by after the total cumulative production of 382 tU. In 1992, the decision for final closure and subsequent decommissioning of the facility was made. In 1994, the plan for decommissioning the centre was accepted by the Slovenian Government Authorities.

The decommissioning plan for the Zirovski vrh production centre provides for the following steps:

- Permanent protection of the biosphere against the consequences of the mining operation. This includes permanent protection of the surface against displacement and subsidence, sealing of the shaft and tunnels against surface waters, airtight sealing of the shaft and tunnels, as well as the provision of an unobstructed run-off for mine waters.

- Permanent remediation of the ore processing plant site is planned in such a way that the remaining facilities can be used for other industrial purposes.

- Permanent rehabilitation of the mine waste and mill tailing areas, including stabilisation of the disposal sites. This involves preventing infiltration of precipitation to the wastes and protecting the surfaces from erosion. Additional measures are being taken to prevent the solution and transport of harmful chemicals into ground and surface waters, and to control contamination by radon.

Remediation of the Zirovski vrh mine site is underway.

• South Africa •

Background

South Africa has a long history of mining activities. Today the mining industry is one of the cornerstones of the country's economy. In 1927, the presence of uranium in gold ores of the Witwatersrand was identified. Lack of any commercial value made it a scientific curiosity until the world-wide search for uranium in the mid-1940s focused attention on the Witwatersrand. The first uranium plant was commissioned in 1952, with production rapidly building up to 1959 when 26 mines were feeding 17 uranium plants to produce 4 954 tonnes uranium. Production declined till the oil crisis in the 1970s when production surged ahead to a peak at over 6 000 tonnes uranium per year in the early 1980s. The collapse of the uranium market led to a decline in production. As a result by 1996 only five uranium plants were operating to produce 1 436 tonnes uranium. An additional plant closed early in 1997, resulting in a further 20% decline in production in 1997.

Since the first production of gold in 1886 a total of 5.6 billion tonnes of ore have been treated and deposited on tailings dams along a 400 kilometre arc along the northwestern margin of the Witwatersrand Basin. Uranium occurs ubiquitously in the gold ore, but not all was treated for uranium because of low grades. In total about 732 million tonnes of the gold ore was treated for uranium, the majority of which was in the West Rand, West Wits Line, Klerksdorp and Welkom goldfields in the western and southwestern part of the Basin. Uranium is produced exclusively as a by-product to gold production with only the Afrikander Lease and Beisa mines being intended primarily for uranium production. However, neither of these facilities were a commercial success. Three processing operations were initiated in the 1970s to extract gold, uranium and pyrite from the mine tailings that were already located on the surface. Two of these facilities ceased production in the late 1980s and early 1990s. The third plant, ERGO, currently produces only gold and pyrite.

The only non-Witwatersrand uranium producer is Palabora located in the Northern Province. At this site uranium is produced as a by-product to copper production from an open-pit carbonatite mining operation. The grades are extremely low, but the large scale of operations makes it economically viable to extract uranium from the heavy minerals concentrates produced.

Legislative framework

Only two Acts specifically control ionising radiation, namely the Nuclear Energy Act 131 of 1993 and the Hazardous Substances Act 15 of 1973. However, reference to radioactive material can be found in other Acts. Those that have a bearing on the mining and milling of uranium are primarily the

Minerals Act 50 of 1991, as amended by the Minerals Amendment Act 103 of 1993 and the Health Act 63 of 1977. Other Acts of relevance to mining activities in general, not specifically to uranium mining, are the Water Act 54 of 1956, the Atmospheric Pollution Prevention Act 45 of 1965, the Conservation of Agricultural Resources Act 43 of 1983, the Environmental Conservation Act 73 of 1989 and the Mine Health and Safety Act 29 of 1996. It should be noted that in the new political situation in South Africa, all Acts are, or will be, undergoing revision. Currently the Nuclear Energy Act, the Minerals Act and the Environmental Conservation Act are under review.

The Hazardous Substances Act has sections whereby radioactive materials can be declared "Group IV hazardous substances" and thus fall under the purview of this Act. However to date the Minister of National Health and Population Development has not utilised this provision; thus the mining and milling of uranium do not have to comply with any statutory requirements under this Act.

The Nuclear Energy Act provides for the establishment of the Atomic Energy Corporation of South Africa Limited (AEC) and the Council of Nuclear Safety (CNS), which regulate the licensing and security of nuclear activities. The AEC does not have any regulatory functions to fulfil with regard to the operations of uranium mines or milling plants. It does provide research, technical assistance and expertise to the mines, in the fields of radiation monitoring, decommissioning and decontamination.

The CNS was initially a consultative body involved in the process of issuing nuclear licenses by the AEC. The amendment of the Nuclear Energy Act in 1988 resulted in the CNS assuming powers of regulation over nuclear installations and activities involving source materials, which had previously been vested with the AEC. Its prime function is to protect the public against the risk of nuclear damage. Pursuant to the risk in regulating source materials, any uranium mining and milling must be licensed by the CNS. The license specifies the terms and conditions under which these activities may be carried out.

A requirement for uranium producers licensed by the CNS is an approved decommissioning and rehabilitation plan specifically for the location concerned. Once the plan is approved it is binding on the company. Department of Mineral and Energy Affairs guidelines require the provision of particular information in the plan, namely:

- A plan indicating all disturbed areas used, showing the type and duration of use.

- A record of radiological measurements and an evaluation of the potential hazards.

- A detailed description of all measures and methods to be used in the rehabilitation process, including details concerning the personnel, plant and material, control and monitoring measures, and use of contractors.

- Expected duration of work.

The licence requirements can only be waived where the CNS has declared in writing that the risk of nuclear damage cannot exceed the limits consistent with health and safety, or the objects of the Act are effectively achieved by the provisions of other applicable legislation.

The Minerals Act makes environmental provisions for mining activities in general, all of which are applicable in the case of uranium mining. In terms of the Act, any prospecting or mining activity must be preceded by the submission of an Environmental Management Programme Report (EMPR) to the Ministry of Mineral and Energy Affairs for approval. This is to ensure that the activity is environmentally acceptable. The examination of the EMPR is co-ordinated by the Ministry, but the Ministries of Agriculture, Environment Affairs, Water and Forestry and Health and Population

Development are actively involved in the process. The EMPR must include a description of the pre-mining environment, a motivation and description of the project, an environmental assessment and an environmental management plan. The EMPR is not intended to be an exhaustive description of the project, but must be comprehensive enough to define all salient features, as the EMPR becomes a binding document on approval. Final closure of the project will only be granted if the EMPR has been adhered to during the duration of the project and its decommissioning.

Operational considerations

The initial stage in any uranium mining and milling operation is the exploration programme which identifies and delineates the orebody to be exploited. These activities have always been controlled by the Minerals Act and its predecessors, and very little legislation was in place in the past to protect the environment. This has changed, as the new Minerals Act and its amendment, requires that any such activity may not proceed without an approved EMPR. The authority for this approval vests with the appropriate regional director of the Department of Mineral and Energy Affairs. To avoid any undue delay, pending approval, the regional director may grant temporary authorisation for the work to commence, subject to any conditions which may be determined by him. It is encouraging to note that the mining companies have been proactive with regard to environmental management and had put programmes in place prior to the implementation of the Act.

Similarly the Minerals Act requires an EMPR prior to commencement of any mining activities, not just uranium mining. Where uranium is to be extracted, a licence must be granted by the CNS which stipulates the conditions under which the mining and processing may take place. Particular reference is made to the protection of society from radiological impacts. The society can be considered to fall into two categories, the occupational work force of the mine and mill, and the public around the mine. Radiological monitoring programmes are put in place in the mine and mill to ensure that exposure of workers to radiation is kept below the statutory limits. Measures are also taken to minimise possible exposure, such as backfilling old stopes to prevent accumulated radon emanations from being circulated into active workings by the ventilation system. In these areas the AEC provides extensive technical assistance and expertise including designing and implementing monitoring and control programmes. The AEC also assists in carrying out radiological hazard assessments.

Exposure of the public to radiation hazards must be assessed for every uranium mining and milling operation. The CNS has published guidelines for such assessments. Exposure may be effected by water-borne or airborne transport. The presence of pyrite in the tailings dams causes lowered pH levels, and thus enhanced leaching of uranium and heavy metals. The low uranium grades in the tailings do not pose a serious problem, but surface runoff is controlled by the construction of dykes and polders, while seepage into the groundwater is minimised by the use of evaporation ponds. Currently the mines are implementing screening studies to evaluate potential radiation hazards to the public. These evaluations must be completed by the end of 1998. These will be followed by establishing and implementing monitoring and control systems.

A third path whereby the public may be exposed to radiation was recently highlighted. Steel scrap resulting from the dismantling of uranium processing facilities or sulphuric acid plants may be contaminated with radioactive materials. The CNS has issued guidelines for the classification and handling of this contaminated scrap material, which must be decontaminated before recycling. The AEC provides a specialist radioactive decontamination service to the mining industry to assist in the processing of this scrap material.

The generation of wind-borne dust and the erosion of the tailings dumps are serious problems which are countered by revegetation programmes aimed at stabilising the dumps. Direct revegetation is the first stage, but the acidity of the dumps poses problems. The use of lime to neutralise the surface acidity, or covering the tailings with a layer of top soil are measures being utilised to encourage the growth of vegetation on the tailings dumps.

Occupational radiation exposure hazards are not confined to uranium mining and milling operations. The presence of monazite in heavy mineral sands and fluorite deposits constitute a potential radiological hazard. Where these deposits are exploited, occupational radiological monitoring and control programmes are in place. Currently monazite is not extracted. However, special measures have been implemented for the safe storage of the monazite-bearing tails in such a way as to facilitate the recovery of monazite if market conditions improve.

There is so far very limited experience in the decommissioning and rehabilitation of uranium mining and processing facilities in South Africa. A number of uranium plants have ceased operations over the years, but the only uranium producing mine which has closed down recently is the Stilfontein mine in the Kleksdorp Goldfield which ceased operation in 1993. The uranium plant is currently being demolished and the entire site is being rehabilitated for reuse. The owner, Gencor, can be regarded as the leading company in South Africa for the rehabilitation of uranium production facilities. Their Stilfontein project will be a pilot project for the dismantling and demolition of uranium mills which will be used to establish guidelines for future projects of this nature in South Africa.

An important aspect of final closure of mining operations are the financial implications. All mines now have rehabilitation trust funds which are funded on an annual basis from operating profits. At the end of the mine's life these accumulated funds will be utilised to effect the environmentally acceptable rehabilitation of the site.

Conclusions

Lack of appreciation of the impact on the environment of mining operations, and uranium mining operations in particular, have in the past resulted in negative effects on the environment. In recent years mining companies have become more environmentally conscious. In fact many companies have implemented environmental standards and programmes which go beyond mere compliance with the statutory requirements now in place under the new Minerals Act and the requirements of the CNS. An important change in focus is the planning of operations so that they are conducted in an environmentally acceptable manner that minimises negative environmental impacts during the production period. Intelligent environmental management has become more prevalent. In addition, mining companies are publicising their environmental concepts.

With regard to uranium mining and milling operations, South Africa has been somewhat behind the developed nations in its environmental consciousness, but this has rapidly changed. Legislation is now in place to control these operations, and although the legislation is in a state of flux with important Acts currently under review for possible change, positive progress has been made. Systems and programmes are in place, or are being implemented to monitor and minimise the radiological and other impacts on the environment. This helps ensure the future well being of society.

• Spain •

REGULATIONS

Law 25/64, dated 29 April 1964, on Nuclear Energy

This law regulates all aspects and activities in the nuclear energy field in relation to nuclear installations, radioactive installations, radioactive mining and materials transport.

Law 15/80, dated 22 April 1980, on the setting up of the Nuclear Safety Council (CSN)

The CSN, which is independent from the State's Central Administration Organisation, is the only one qualified to deal with radiological protection and nuclear safety issues. The functions of the CSN are to:

- Propose regulations to the Government in nuclear safety and radiological protection issues.
- Issue to Industry and the Ministry of Energy the reports on licences for construction start-up, operation and closure authorisations of the nuclear installations and radioactive installations.
- Perform any kind of inspection.
- Collaborate in the emergency plans.
- Monitor and survey radiation levels in the installations and in surrounding areas.
- Issue the required licences for the operation personnel.
- Advise public administration courts and institutions.
- Maintain official relations with similar foreign organisations.
- Inform the public.

Decree 2869/72, dated 21 July 1972

By this Decree nuclear energy and radioactive nuclear facilities regulations are approved. These regulations determine administrative licences, installations, testing and start up, inspections, personnel and documentation. Uranium ore processing plants are classified as first category radioactive installations, and authorisations are required for their siting, construction, start up, operation and closure.

Royal Decree 53/1992, dated 22 January 1992

By this decree the health protection regulation concerning ionising radiations is approved. The purpose of this regulation is to establish the basic standards for radiological protection to prevent non-stochastic biological effects and limit likelihood of the stochastic biological effects to levels considered acceptable for workers professionally exposed, as well as the general public, as a consequence of activities implying exposure risk to ionising radiations.

Location of mining zones in mainland Spain

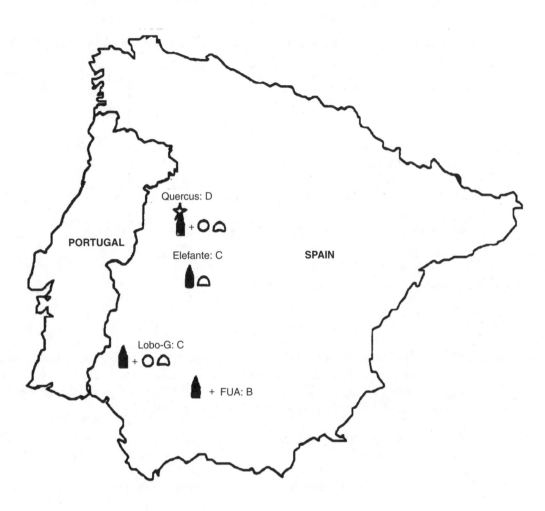

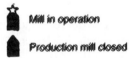 Mill in operation

Production mill closed

B: No production; site remediation completed
C: No production; site remediation being implemented
D: In operation; site remediation being implemented

○ Open-pit mine

◠ Heap leaching

+ Tailings storage

Euratom Directives No. 80/836 and No. 84/467 were incorporated into the Spanish legislation through the publication of the Health Protection against Ionising Radiations Regulation.

Safety guidelines applicable to the uranium mining and milling activities published by the CSN

GS-5.6	Requirements for obtaining and using licences for personnel in radioactive installations	ISBN 84-87275-30-3
GS-5.8	Guidelines for elaborating information regarding radioactive installation exploitation	ISBN 84-87275-24-9
GS-7.2	Requirements for obtaining classification as an expert in protection against ionising radiations and to qualify for being responsible for the corresponding Technical Unit or Service	ISBN 84-87275-29-X
GS-7.3	Guidelines for establishing Protection Techniques Services or Units against Ionising Radiations	ISBN 84-87275-23-0
GS-7.4	(Revision 1) Medical Care Guidelines for workers, exposed to ionising radiations (surveillance)	ISBN 84-87275-22-2
GS-7.6	Manuals outlining the radiological protection for installations of the nuclear fuel cycle	ISBN 84-87275-49-4

Decommissioning

Spanish regulations require that the owner of the facility, prior to finalising production activities, present a decommissioning plan ("Plan de Clausura") to the Ministry of Industry and Energy (MINER). This plan is evaluated by the Nuclear Safety Council (CSN) before authorisation is granted to permit the owner to undertake dismantling and decommissioning operations.

The decommissioning plan is to be based on instructions received by the owner from the CSN and comprises the following sections:

- Operation history of the facility describing its characteristics and processes, and including inventory of installations, waste rock and tailings affected by the decommissioning.

- Decommissioning criteria analysing the different alternatives considered and corrective actions to be implemented in relation to land decontamination, materials declassification, protection from barrier degradation and human intrusion.

- Site characterisation from the radiological point of view, paying special attention to the area's natural background radiation, which will serve as the reference for achieving decommissioning objectives. Evaluation of characteristic parameters related to meteorology, hydrology, hydrogeology, water and soil use factors, geology, seismicity and geotechnics, demography, as well as socio-economic factors.

- Restoration Works in which the technical bases for the design of monitoring and treatment of liquid wastes, solid materials and wastes management are established. In addition, the design of covers and stability analyses of the decommissioned structures and final reforestation of the affected areas are specified.

- Analysis of the Radiological Impact on workers and the public, both during the operations of the decommissioning phase as well as the final impact to the public. Once decommissioning operations have been concluded, the radiation sources and different exposure pathways are to be evaluated.

- Quality Control Procedures ensuring that operations are carried out in compliance with established criteria. These procedures include aspects such as organisation, control and design, verification, operating instructions and procedures, document and archives control, inspections and audits.

- Post-decommissioning Monitoring Programme of the radiological condition of the site. This is to be implemented during a final verification period (from minimum five years to a maximum of ten years).

Remediation of residue disposal sites

In Spain the situation at radioactive installations related to the mining and milling of uranium ores can be summarised in the following manner:

- Quercus Plant, located in Salamanca includes an open-pit mine, milling plant, static-leaching heaps and tailings ponds. It is currently in operation.

- Lobo-G Plant, located in Badajoz, includes an open-pit mine, milling plant and tailings ponds. It was closed down in 1997.

- The Uranium Processing Plant in Andujar (F.U.A.), located in Jaén, included a milling plant and tailings ponds, but has been closed down.

There are also several small uranium mines which have been abandoned in which mining operations were carried out during the 1960s and 1970s. They are included in the Administration's General Remediation Plan. A contract to carry out the restoration of nineteen of these sites was awarded to Empresa Nacional del Uranium, S.A. (ENUSA) in September 1997. The sites are in the Andalucía and western Extremadura regions and the estimated time to complete the work is two years. The activities were started in November 1997.

Decommissioning activities during 1996

A ten-year supervision programme, which was designed prior to the announcement of the closure of the old Andujar uranium concentrate plant (which operated between November 1959 and July 1981), was implemented following completion of restoration. The dismantling and restoration activities were completed in June 1994. This work was performed by Empresa Nacional de Residuos Radioactivos, S.A. (ENRESA).

Decommissioning of the waste dump and processing plant, at ENUSA's La Haba Centre in Badajoz province, was authorised in November 1995. Dismantling and restoration of the site was carried out during 1996 and was completed in 1997. Materials obtained have been placed in the tailings dam that has been capped with a 3-8 m thick layer of high-clay content waste material from the mine. The open pit was decommissioned during 1995, including revegetation.

After completion of the dismantling and closure operations at La Haba, a five-year supervision programme will be established to verify the fulfilment of the design and construction criteria imposed by the Spanish Nuclear Safety Council.

At the Saelices Mining Centre (Salamanca province), a project for decommissioning the old Elefante treatment plant, together with the heaps from the old heap leaching operation, was presented to the Nuclear Safety Council at the end of 1996. The authorisation is pending.

The Nuclear Safety Council and the Ministry of Industry and Energy have approved a plan for the restoration of nineteen old uranium mines operated by the former Junta de Energía Nuclear (JEN) between the early 1950s and 1981. They were active either as test or production mines, and the ore was treated at the Andujar Plant. Of the 19 mines, 13 are in the Extremadura autonomous region and 6 in Andalucía. In March 1997, the Extremadura autonomous government approved the decommissioning project. The work was started in November 1997.

Environmental control of operating mining centres

Presently there is only one active uranium mining centre in Spain, at Mina Fe near Ciudad Rodrigo (Salamanca province).

The Fe mine is an open pit operation owned by ENUSA. Crushing and milling of the ore is done close to the pit, followed by both conventional and heap leaching processing to recover the ore. The environmental control involves airborne effluents; liquid effluents after neutralisation; deposition of tailings in a dam and soil restoration to control erosion.

The implementation of the control operations is carried out by ENUSA with its own organisation and laboratories. Follow-up and supervision of these control operations, is carried out by the CSN that reports regularly to the Spanish Congress and Senate.

• Sweden •

Life-cycle studies of electricity generation

In Sweden, Vattenfall has carried out life-cycle studies for different methods of electricity production: hydro power, nuclear power, oil condensing, gas turbines, biofuelled combined heat and power plants, wind power and natural gas combined cycle.

The study included the building of power stations as well as the fabrication of building materials (such as steel, copper, aluminium, lead, concrete etc.), the production and refining of fuel, operation of the plants, reinvestments, modernisation and decommissioning. All emissions are summed up and reported, regardless of whether they take place locally, regionally or globally.

In the nuclear power case, emissions from uranium mining, conversion, enrichment, fuel fabrication, as well as spent fuel handling and final storage were summed up together with emissions from building and decommissioning of the nuclear power plants.

Many environmental impacts were studied, such as emissions of carbon dioxide, sulphur dioxide, nitrogen oxides, radioactivity, energy needs, land needs, and biodiversity, etc.

One of the results of the life-cycle studies is that emissions of carbon dioxide, sulphur dioxide and nitrogen oxides are very low for three methods of producing electricity: hydro power, nuclear power (including the fuel cycle) and wind power. In contrast all of these emissions are high for oil condensing and gas turbines. Carbon dioxide and nitrogen oxides emissions are high for natural gas power; and sulphur dioxide and nitrogen oxides are relatively high for biofuelled combined heat and power plants.

Restoration of the closed-down uranium mine in Ranstad, Sweden

The uranium mine in Ranstad was opened as part of the Swedish Nuclear Power Programme. Uranium was extracted by percolation leaching of alum shale using sulphuric acid. The shale has a grade of about 300 g/tU and also contains about 22% of organic matter and about 15% of pyrite. The plant operated from 1965 to 1969. During this period, 1.5 million tons of alum shale were mined by open pit and 215 tU were produced. The total amount of the tailings is about 1 million cubic meters, containing about 100 tons of uranium and 5×10^{12} Bq of radium 226. The tailings cover an area of 250 000 m^2. The open pit mine was a trench 2 000 m long by 100-200 m wide and 15 m deep.

As the licence for operating the Ranstad plant expired in 1984 discussions about the final restoration started. At that time the mill tailings were covered by a 0.5 m thick layer of moraine material. Leachate from the tailings was collected and purified before it was released to the Flian River. The environmental consequences were not critical but the continuous pumping and treatment of leakage water was costly. Consequently, a plan for final restoration had to be prepared. The open pit was kept dry until November 1991 when pumping was stopped.

Studsvik AB was responsible for the project planning, sub-contracting, performance of the restoration, as well as the environmental monitoring. Since 1992 AB SVAFO, owned by the Swedish nuclear power industry, has been responsible for the restoration.

Objectives

The objective of the restoration plan for the Ranstad site is to make all future maintenance unnecessary. When it is fully completed it is anticipated that the area will comply with environmental legislation without any further intervention.

Restoration

Studies were performed during the 1970s and 1980s to find effective ways for controlling and reducing the leaching processes. The aim of the investigations was to find methods for reducing the infiltration of moisture from precipitation, as well as prevent oxygen from entering the tailings. One of the activities undertaken to investigate this problem was constructing a test pile, consisting of 15 000 t of mill tailings, in 1972. This was used for studying the effectiveness of different cover systems. In 1985, the general planning of the restoration was started with investigations including collection of available information and completion of maps of geology, hydrology and water quality. A detailed plan for the restoration was submitted to the County Administration in October 1988. After review, the authorities granted permission starting the project in January 1990. The authorities required that a release and recipient control programme for the concentration of uranium and radium, as well as heavy metals (e.g. cadmium, copper and nickel) should be performed.

The mill tailings

Establishing a tight cover over the leach residues and additional mill tailings was the most complicated part of the restoration, as the County administration required that the cover has a hydraulic conductivity of less than $5\text{-}10^{-9}$ m/s. Such factors as availability of material, the long-term stability and the experience from using the material also had to be considered in designing the cover system. The preferred cover for the tailings is: 0-0.3 m moraine material in the already installed cover, followed in sequence by a 0.2 m clay-moraine mixture, 0.2 m crushed limestone, 1.2 m moraine and 0.2 m soil-moraine mixture on top. Moraine is a reliable, natural material, having been used in many Swedish hydro-power dam construction projects. Within the disposal site at Ranstad, a sufficient quantity (about 50 000 m^3) of an especially favourable type of moraine containing a fine fraction of clay shale particles was found. This material gives the required low permeability, without the additional step of adding and mixing bentonite. Large quantities of the moraine required for constructing the cover were found both within the disposal area and in the open pit.

The quality of the sealing layer was strictly controlled by making laboratory and field tests of such factors as the moisture content, fine particle fraction and compaction. The cover had to be installed over about 250 000 m^2. After testing indicated the quality of the first layer met the requirements, the limestone layer for drainage and the moraine layer for protection were successively installed. For monitoring the effectiveness of the cover system, a large number of observation wells was installed to monitor the groundwater level over the impervious, lower layer. For monitoring the diffusion of oxygen, as well as the infiltration of percolating water, lysimeters have been installed beneath the tight, lower layer.

The open pit mine

At the open pit mine the very impervious alum shale open pit floor has been used to create a lake with an area of 250 000 m^2. The groundwater filled the mine area to about the original level and it discharged to the environment. Before filling the pit with water, waste rock was moved from near the mining front to the bottom of the pit and covered by moraine. In total over 0.5 million m^3 of rock and earth material was moved before water filled the pit. The filling of the lake volume with 1.3 million m^3 of water was completed by mid-1993.

Results

The aim of restoration at the Ranstad site is to reduce the effects of the mining activities to levels that are acceptable compared to the background values of the surroundings. A comprehensive environmental control programme has been implemented to monitor water quality within the disposal area, as well as in the former open-pit area and their recipients. Standard parameters such as macro constituents, pH, alkalinity, as well as heavy metals and uranium and radium are measured. Discharge measurements are also performed to estimate the transport of different elements, i.e. the total load of these elements.

The primary restoration was finished at the end of 1992, when installation of the tailings cover-system was completed. It can be concluded, three years after completion of the primary restoration, that the concentration of pollutants in the leachate from the disposal area has decreased to about one third of levels prior to restoration. Furthermore, the concentrations of uranium and heavy metals, such as nickel, in Lake Tranebärssjön (the former open-pit) have stabilised at a relatively high level. The difference in the concentration of these metals between the surface and bottom water is marked. The stratification in the lake is very stable with anaerobic conditions below 10 meters depth.

Conclusions

Predictions of the environmental consequences were an important part of the investigation phase of the restoration of the Ranstad site. They gave a scientific base for judgement of the restoration efforts and the possibility for the responsible authorities to evaluate the restoration plan. The techniques for preparing a tailings cover barrier with a low hydraulic conductivity, using an appropriate quality assurance and control programme, have been well tested and established.

Following completion of the primary restoration programme, the project is in a follow-up phase. Three years after the application of the cover-system the concentration of heavy metals (including uranium) in the leachate from the mill tailings is decreasing. However the decrease is slow and it will take at least another three-year period before the metal concentrations are in accordance with acceptable levels when compared to background values.

In the stratified Lake Tranebärssjön, located in the former open-pit mine, the concentrations of most metals increased initially in the bottom water. However, the concentrations in the surface water for such elements as uranium, nickel, cadmium and cobalt are stable at a relatively high level.

The requirements of the tailings cover-system is to minimise the infiltration of both oxygen and moisture from precipitation into the tailings. Observations indicate that the infiltration is about 10-15% of the precipitation and that the content of oxygen in the tailings is less than 1%. Thus the cover-system is fulfilling the requirements.

• Ukraine •

The Ukrainian uranium industry started with exploration for commercial deposits in the Ukrainian provinces of Dniprovska and Kirovogradska in 1944. Now, there are more than 20 commercial uranium deposits. The most important ones are Vatutinske, Severinske and Michurinske. Uranium production in Ukraine started by using an existing iron production centre at Zholtaya river. The settlement developed into the town of Zheltiye Vody.

The uranium and nuclear fuel industries have two large production centres: the eastern mining concentration mill in Zheltiye Vody (VostGOK) and the industrial amalgamation Pridnieprovskiy Chemical Plant (P.Ch. Z.)

VostGOK currently operates the underground mines situated near Kirovograd and Smolino in the Kirovogradska province. From 1947 to 1991, P.Ch.Z., situated in the Dnieprovska province, was an uranium production centre. It is now operating with different nuclear metals, materials and chemicals.

Both VostGOK and P.Ch.Z. are situated in the central part of Ukraine. The area has a developed infrastructure, high density of population, fertile soil and intensive agricultural activity.

Description of the production and waste

Industrial site in Zheltiye Vody

The industrial site in Zheltiye Vody is VostGOK's main centre of operations (see Figure 1). Both mining and supporting companies are located at this site. They are: the Olhovskaya and Novaya mines, a hydrometallurgical plant (HMZ), sulphuric acid plant (SKZ), energy facilities and the company office.

During operation of the Zholtovodskiy mine several facilities were developed. They include: the Gabayevskiy and Veseloivanovskiy quarries, 3 tailings dumps designated KBZh, Sch and R, and one open pit mine. Mining at Olhovskaya and Novaya accumulated 550 000 m^3 of dumps.

The area of effected land consists of 968 hectares including dumps (19.1), open pits (17), quarries (50.6), tailings (645.6) and others.

The Olhovskaya Mine was closed in 1980.

The mud and rock waste from the Novaya Mine have been moved into the open pit. An old iron-ore quarry was used as the site for the KBZh tailings impoundment. These tailings are situated within the sanitarian-protective zone of hydrometallurgical plant, three kilometres from the town. The quarry was filled in 1987. The total quantity of tailings is 19.34 million tonnes.

Most of these tailings have been rehabilitated. The remaining area is used as an emergency reserve place for tailings in the event of an emergency at the hydrometallurgical plant. The tailings, consisting of leached uranium ore, contain 0.007% uranium. The radon exhalation from the tailings surface fluctuates between 0.05 and 3.0 Bq/m^2 per second.

The Sch tailing site includes 250 hectares situated 1.5 kilometre from the town in a ravine called Scherbatovska. Tailings disposal started here in 1979. The uranium content of the tailings is 0.007%. The radon exhalation from the tailings fluctuates between 0.5 and 2 Bq/m^2 per second. The total alpha-activity of the tailings is 1.8×10^{15} Bq.

The R tailings, consisting of 230 hectares, are located not far from the Yellow River in Rabery ravine. This tailings area was used for iron ore tails from the Novaya Mine since 1969.

Annually, 2.4 million m^3 of mine water are decontaminated by removing radionuclides before the water is released to the environment.

Industrial site of Ingulsky mine

This site is situated south of Kirivograd (see Figure 2). The uranium deposit, situated at 700 m depth, was mined by three shafts. After preliminary sorting, the ore is divided into commercial ore, marginal ore and mud/waste rock.

The marginal ore and mud/waste rock are transported to the dumps.

Figure 1.　**VostGOK. Scheme of the Zhovty Vody industrial site**

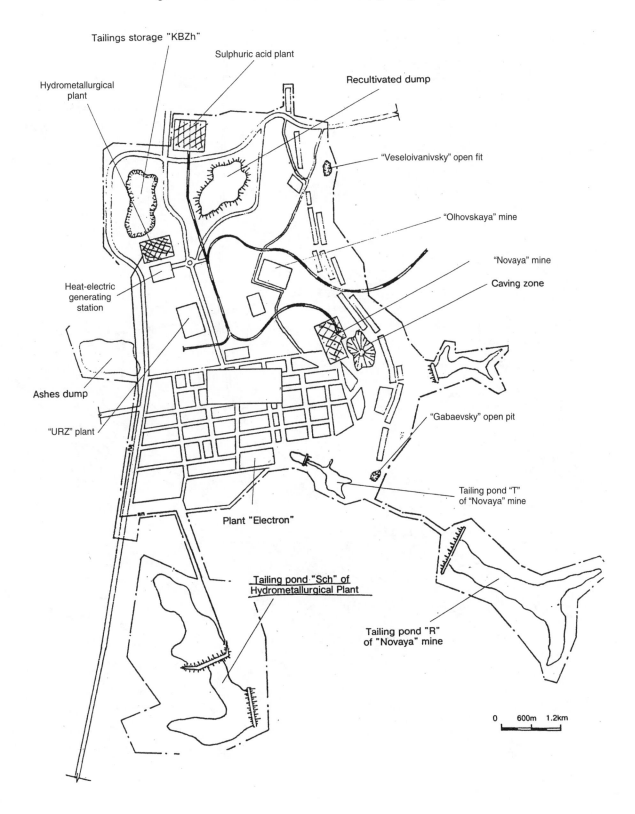

Figure 2.   **VostGOK – Ingulsky mine group – Scheme of industrial site**

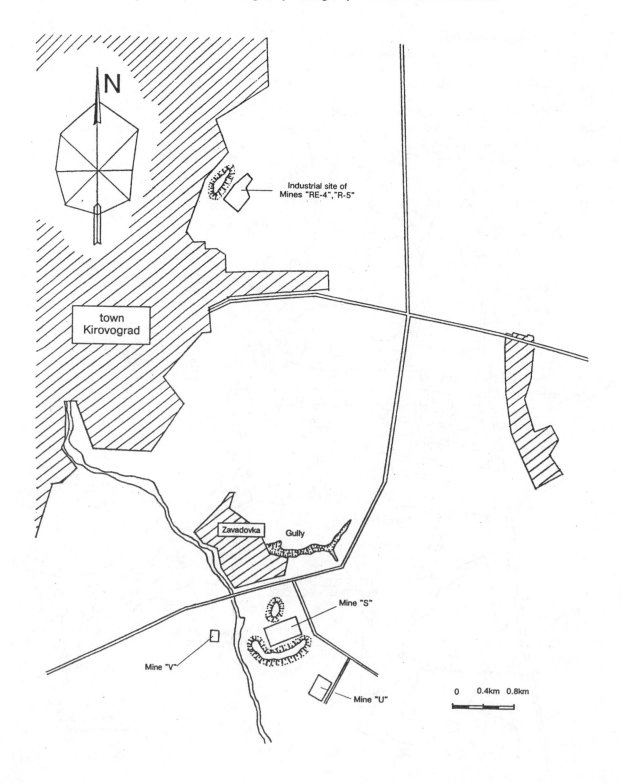

Land rehabilitation

The following land rehabilitation projects are being carried out in Ukraine: recultivation of the industrial sites of uranium deposits; rehabilitation of the Olhovskaya mine and shaft and related territory; KBZh tailing rehabilitation; and decontamination of the Lviv railway cargo station in Muchachevo.

Ukraine is also proceeding with other projects such as the recultivation of the dumps and tailing in Mayly Say and the rehabilitation of the Shecaftar mine, both located in Kirgyztan; the decontamination of the Brest cargo station in Belarus; and rehabilitation of the Ingulsky mine, Ukraine.

Legislative framework

Since the 1991 Independence Act Declaration, Ukraine has been changing the laws of the former USSR and passing new national laws and bills. This includes regulations for geology, mining and energy activity. The new Constitution of Ukraine was declared in 1996. At present, there are new legislative bases to carry out rehabilitation projects:

- The Law of Ukraine regarding nuclear energy utilisation and radiation safety.

- The Law of Ukraine regarding radioactive waste management.

- The Bill of Ukraine regarding clean-up regulations upon liquidation/closing, redirecting the direction of industrial activity and temporary closure of mining and processing companies operating with radioactive ore.

Rehabilitation of the Olhovskaya mine and shaft and related territory

The rehabilitation of the Olhovskaya mine and shaft and related territory was carried out from 1979 to 1982. The technology used included selective removal of the waste heap and waste rock dumps and marginal ore. Part of this waste was transported to the VostGOK hydro-metallurgical plant, while another was moved to the tailings dam. However, the largest part of this rock was dumped in the nearby quarry. A total of 550 thousand m^3 of mining rock was moved and/or utilised.

The contaminated soil from below the dumps was excavated to a depth of 1 metre and dumped in the quarry. The recultivated area of 15 hectares has a gamma-radiation of 0.10-0.22 microSv/hour. The rehabilitated area is used for civilian and industrial buildings and agricultural activity.

Recultivation of the KBZh tailings

Recultivation of the KBZh tailings started in 1991 and is continuing. About 85% of the territory of the KBZh site has been covered by a 0.4 m thick clay loam cap to protect against dust migration. The project requirements for the protective covering envisages the following structure: 0.4 m of loamy clay, (which is already in place), 0.4 m of rock, 3.5 m of packed loamy clay, and 0.4 m of black earth. After the final rehabilitation, this 55 hectare territory will be returned to agricultural use as a pasture.

Recultivation of the Ingulsky mine dumps

To reduce the negative influence on the environment at the Ingulsky mine dumps, waste rock from mining will be dumped into ravines. A total 500 thousand m³ of mining waste, with an activity of 370 to 39 000 Bq/kg, will be relocated. At the bottom an impermeable 0.4 m thick clay layer will be placed. Capping the waste will include: a 0.4 m clay layer, a broken stone drainage layer, a 1.5 m loam hydro-isolation layer and on top a 0.3 m layer of fertile soil.

According to the plan, the radon exhalation from the site will be the same as background level and the underground waters will not be influenced for thousands of years. About 7.4 hectares of the rehabilitated territory will be planted with trees.

Rehabilitation problems of the industrial sites in the town of Zheltiye Vody

At VostGOK's industrial sites, radioactive materials were used for constructing dwellings and apartment houses between 1940 and 1970. In the 1980s, VostGOK decontaminated the most polluted parts of the town. However, much work remains to be done. The gamma radiation and radon concentration in buildings in the town are shown in Figure 3. In 1994, 2 000 measurements were carried out to assess the indoor radon levels in the town.

On 8 July 1995, the Council of Ministers of Ukraine decided on directions about the "State Measures Programme on Radiation Safety and Social Maintenance of Zheltiye Vody Inhabitants in Dniepropetrovska Province".

The first 10-year stage of the programme envisages the following measures: cleaning/liquidation of areas of contaminated territory in the town; taking action to lower the indoors radon concentrations in houses; and elaboration and implementation of radio-ecological monitoring.

State programme for improvement of radiation protection on the sites/facilities of the atomic industry of Ukraine

The main units of the state planning programme for improvement of radiation safety in Ukraine are shown in Figure 4. The programme covers all identified industrial sites and environmental issues in uranium mining and milling. The implementation cost of the programme is USD 360 million.

Planning for further technology development of uranium mining and milling

The uranium board of Ukraine and its engineering group focus on conducting industrial mining and milling research to develop the technology for the modelling of the ecology system to minimise the negative environment impacts.

Innovations such as those listed below are preconditions for this step:

- P.Ch. Z. (Managing Director, Mr Y.F. Korovin) produces a very effective ion-exchange resin for the high-level uranium extraction from uranium ore.

- VostGOK (Managing Director, Mr M.I. Babach) has developed an advanced milling technology and is studying a new technology of pulp utilisation in underground mines.

144

Figure 3.   **Gamma Radiation and Radon Concentration**

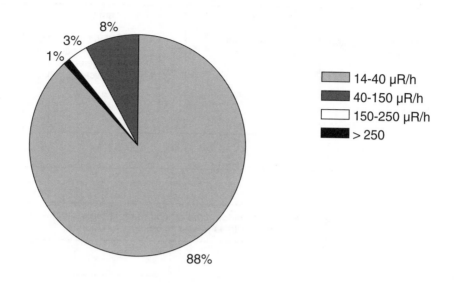

**Gamma radiation dose rate on studied territory
of the town Zhovty Vody**

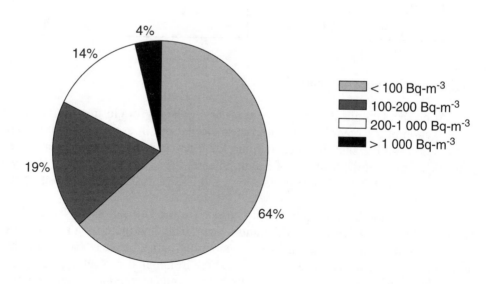

**Equilibrium equivalent radon concentration in buildings
of the town Zhovty Vody**

Figure 4.   **The main units of the planning State Programme for improvement of radiation protection**

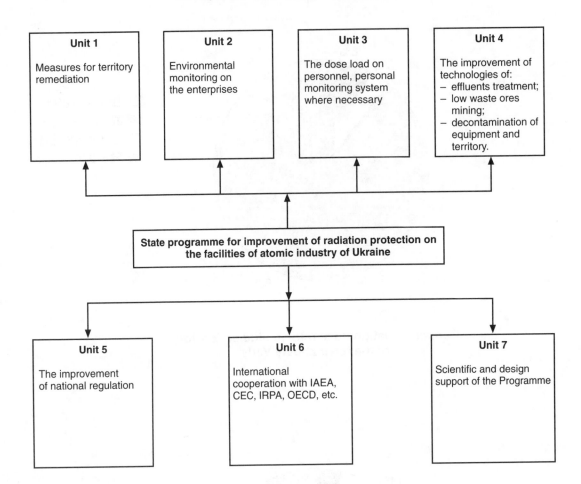

Legislation for mine rehabilitation

At present Ukraine does not have a law or regulation for closure of uranium mines and mills. The investigation of the environmental impact is carried out in association with uranium exploration and exploitation and the shut-down of uranium facilities. It is carried out under decrees of the Cabinet of Ministers, orders of the Ministry of Environmental Protection and Nuclear Safety and requirements of the State Committee of Ukraine for Geology and Utilisation of Mineral Resources.

At present the State Committee of Ukraine for the Utilisation of Nuclear Power has developed a law on "Mining and Treatment of Uranium Ore". This law includes a regulation of relations between mining and treatment of uranium ores; maintaining parity between required mining and treatment of uranium ores with human safety aspects and environmental protection; and regulation of activity in this field, radioactive methods and matters, government regulation and safety control at uranium mining and treatment facilities.

The main terms, definitions and tasks of legal regulation in the field of uranium mining and treatment are formulated in this law. It includes applicable legislation: Constitution of Ukraine, Land Code, such Ukrainian laws as "Utilisation of Nuclear Power and Radioactive Safety", "Environmental Protection", and "Labour Protection".

The law as stated specifies the Government policy and main principles in the field of uranium mining and treating, ownership of uranium facilities, rights of people and their unions in the field of uranium mining and treating, role of government and management bodies.

The legal status of uranium facilities and their staff, as well as the legal regime of territorial distribution of uranium facilities, and responsibility for violation of legislation in the field of uranium mining and treating of ores are also indicated in this law.

Research

Before the law is approved, specialists must carry out geological and ecological research of mines and adjacent territories. This work is done during exploration, exploitation and decommissioning or conversion of uranium facilities, or when the uranium deposit is mined out.

This research includes measuring background environmental parameters; determining the radiation levels within the deposit (based on radiometric and emanation surveys conducted at a scale of 1: 5000); characterising surface and underground waters, and any changes of the waters during exploration and exploitation; characterising the soil by conducting radio-ecological and toxicological test on a grid of 400 m x 200 m in the deposit and the adjacent territory within a distance of 1 000 m from deposit borders; and quantitative and qualitative characterising of mine water.

At present Kirovgeology is conducting radon monitoring of the environment around uranium mining and milling facilities. There are some problems regarding this task particularly deciding which method is more suitable – passive or active. The understanding is that it is necessary to develop a National Radon Programme for the country.

Ukraine is ready to carry out international co–operation concerning environmental issues in uranium mining and milling related to experience in this field.

• United States •

LAWS AND REGULATIONS

Between 1975 and 1978, the Department of Energy (DOE) and the Energy Research and Development Administration (ERDA), successor agencies to the Atomic Energy Commission (AEC), completed studies of uranium mill sites. The studies examined mill sites that had produced uranium concentrates for the AEC, had subsequently ceased operations, and were considered "inactive". The agencies determined that uranium mill tailings located at these "inactive" uranium mill sites posed potentially significant health hazards to the public and that a programme should be developed to ensure proper stabilisation or disposal of these tailings, to prevent or minimise radon diffusion into the environment and other related hazards.

In November 1978, as a result of these studies, the Congress enacted Public Law 95-804, the Uranium Mill Tailings Radiation Control Act (UMTRCA). The law provides for the clean-up and stabilisation of uranium mill tailings at currently inactive uranium-processing sites. UMTRCA vests the U.S. Environmental Protection Agency (EPA) with the overall responsibility for establishing environmental standards and guidelines, but regulatory responsibility for U.S. nuclear facilities (including uranium mills), remains with the Nuclear Regulatory Commission (NRC), or with the "Agreement States". An "Agreement State" is a State that is or has been a party to a discontinuance agreement with NRC under Section 274 of the Atomic Energy Act (42 U.S.C. 2021) and that issues pursuant to a counterpart in its State laws to Section 62 or 81 of the Atomic Energy Act. Nevertheless, UMTRCA is the basis for present-day control of uranium mill sites.

Applicable EPA standards under UMTRCA are contained in the "Health and Environmental Protection Standards for Uranium and Uranium Mill Tailings" (40 CFR Part 192). Licensing and enforcing the regulations, controlled by NRC (or the agreement state), are covered under the "Domestic Licensing of Source Material" (10 CFR Part 40). UMTRCA requires that each license contains provisions regarding decontaminating, decommissioning and reclaiming the licensed facility. The licensee must also comply with supplemental standards, such as "Health and Environmental Protection Standards for Uranium and Uranium Mill Tailings" (40 CFR Part 192).

UMTRCA, however, does not provide for reimbursement of remedial action costs incurred at "active" (at the time of UMTRCA enactment) uranium processing sites containing piles of commingled tailings generated under Federal Government contract and for commercial entities. Two reports were prepared subsequently for Congress on this issue, one by the DOE in January 1979 and one by the General Accounting Office in February 1979. Both concluded that Government assistance should be provided to licensees at these sites.

Title X of the Energy Policy Act of 1992 (Public Law 102-486) established the authority and framework for providing Federal assistance. The DOE is now required to reimburse licensees of "active" uranium or thorium processing sites for remedial action costs attributable to by-product (mill tailings) generated as an incidence of uranium or thorium concentrate sales to the United States. The DOE issued a final rule on 23 May 1994, under the "Reimbursement for Costs of Remedial Action at Active Uranium and Thorium Processing Sites" (10 CFR Part 765). Title X sets a maximum ceiling on individual and total reimbursement. Uranium licensees may receive up to USD 5.50 per dry short ton of Federal-related tailings. Total reimbursement is restricted to USD 270 million for the 13 eligible uranium licensees and USD 40 million to the eligible thorium licensee. Reimbursable activities include, but are not limited to: groundwater restoration, treatment of contaminated soil, disposal of process wastes, removal actions, air pollution abatement, mill and equipment decommissioning, site monitoring, administrative expenses, and other activities necessary to comply with the UMTRCA.

URANIUM MINE RECLAMATION

The NRC does not interpret its authority on environmental remediation as extending to uranium mines. The enforcement of uranium mines is carried out by the individual States by treating uranium mines like any other type of mines. Most States have laws for abandoned mines and regulate reclamation work. Reclamation of uranium mines may be also subject to rules and regulations governed by Department of the Interior´s Office of Surface Mining Reclamation and Enforcement, or by the Bureau of Land Management, in compliance with "Surface Management" (43 CFR Part 3809), provided under "The Federal Land Policy and Management Act of 1976" (Public Law 94-579).

STATUS OF MILLS AND ISL PLANTS

There are a total of 26 uranium mills which produced uranium for either commercial or for both Government and commercial (commingled) purposes. Six of these are on standby and 20 are undergoing various stages of decommissioning. One out of the six uranium mills on standby was partially operational during 1995, producing concentrates from stockpiled ore mined prior to 1993. The Government portion of remedial action costs for the 13 (commingled) uranium mill sites out of 20 sites and one thorium site are reimbursable by the Government under Title X programme.

In addition, there are 24 "inactive" uranium mill sites that had operated for the Government prior to 1970, and associated vicinity properties, located in 10 States, including two mills on Indian tribal lands. Remedial action expenses for these sites are born by the Government (including 10 percent from affected States) under the Uranium Mill Tailings Remedial Action Programme (UMTRAP). Environmental restoration costs for some of these earlier site are generally higher, because the remediation of an abandoned waste pit or tailings pile often requires the exhumation and relocation of all materials previously disposed. Moreover, detailed information on the volume and types of waste that might be generated and disposed are sparse, thus making it difficult to assess the project magnitude and make reliable technical projections.

Seven non-conventional uranium producing plants, consisting of five in situ leach (ISL) and two phosphate by-product (PB) plants, were in operation at the end of 1995. Eight other non-conventional plants (six ISL and two BP plants) were inactive at the end of 1995. The two BP plants were closed indefinitely, and one ISL plant was subsequently restarted in 1996.

• Uzbekistan •

LAWS AND REGULATORY AUTHORITIES

A project to develop a law about "radiation protection of the population" is currently underway in Uzbekistan. Adequate infrastructure and government bodies are envisaged in this project. The bodies will co-ordinate and regulate activity in the field of radiation protection and proper treatment of radioactive waste materials.

The regulation and supervision of radiation safety for the population in Uzbekistan is conducted by:

- The State Committee of Uzbekistan for Nature Protection which co-ordinates all environmental questions in Uzbekistan with other authorities.

- The Uzbekistan State Committee on Safety in Industry and Mining which supervises the Navoi Mining and Metallurgical Complex (NMMC) and investigates the impacts of uranium mining.

- The State Sanitary Ecological Service in the Ministry of Health which supervises radiation protection of the personnel of NMMC.

These regulating bodies monitor the country to assure that rules and norms of radiation safety are observed.

The State Geological Company "Kyzyltepa Geology" has scientific and technical experts in the environmental, geological and hydrological fields. The governmental authorities often request that the experts of Kyzyltepa Geology carry out independent control and monitoring.

Uranium production by Navoi Mining and Metallurgical Complex (NMMC)

The Navoi Mining and Metallurgical Complex (NMMC) is the company solely responsible for mining and milling of uranium in Uzbekistan. This company is a member of the State Concern "Kyzylkumredmetzolo". Today the extraction of uranium from all deposits is only being done using in situ leaching (ISL) technology. However, until recently, uranium extraction was done using open pit and underground mining, followed by conventional milling.

The radiation protection of the population that lives in the mining areas, together with environmental protection, is provided by the operational organisations of the NMMC.

The structure of the NMMC uranium industry consists of:

- Hydrometallurgical plant No. 1 (GMZ-1), which processes uranium raw materials and associated tailings, situated near the city of Navoi.

- The mining companies (Northern Ore Department, Uchkuduk; Eastern Ore Department, Zasafshan; Southern Ore Department, Nurabad; and Ore Department No. 2 in p.g.t. Krasnogorsk).

GMZ-1 has the responsibility of producing uranium concentrates from ores and pre-concentrated solutions collected using ISL from the Uchkuduk, Sabyrsai, Sugraly and other deposits.

Tailings impoundment and waste handling

Wastes from processing of uranium ores are deposited into the tailings impoundment located 5 km west from GMZ-1.

The tailings impoundment which holds the tails from earlier milling of ore mined underground or in open pits, is the most ecologically stressed area of NMMC because of its scale and location in the densely populated valley of the Zarafshan river.

The area of the tailings is 600 hectares. The height of the dam reaches 15 metres. The tailings impoundment has been used since 1964. Now, 59.7 million tonnes of production wastes are in the tailings impoundment. The total quantity of radionuclides in the tailings is about 160×10^3 curies. The tailings and associated radiation are given in the following table.

Zones of tailings	MED of gamma radiation microröntgen/h	Specific alpha-activity (kBq/kg)	Concentration of radon and DPR in atmosphere near earth, (Bq/m^3)	Concentration of long-lived alpha-nuclides, (curie/l)
Surface of worked out maps	300-1 200	80 / 8.0-190	6.0-33	1×10^{-15}
Surface of dams	100-500	–	–	–
Guarding zone (100-110 metres)	100-150	19 / 1.8-39	5.0-30	0.25×10^{-15}
Sanitary protection zone (800 metres)	13-30	37 / 1.1-6.0	–	0.2×10^{-15}
Observance zone	13-18	0.37 / 0.15-1.0	4-13 / DKB-37	0.14×10^{-15} / DKB-0.33×10^{-15}

To lower the ecological stress in the area of the GMZ-1 impoundment, the following measures have been taken:

- A drainage (curtain) system is in operation. This system consists of 24 pumped boreholes drilled to intercept the movement of underground waters moving from the impoundment to the Zarafshan River. This system is designed to recover radionuclide bearing solutions escaping from the tailings pond and return them to the hydrometallurgical plant.

- Sub-surface waters near the tailings are regularly monitored using 108 boreholes drilled for this purpose.

- 38 cottages were built to relocate families of the village of Turkman. Expansion of the tailings pond was located in the sanitary zone.

The radiation situation of NMMC's plants and facilities where mining is carried out is presented in the following table:

Area and name	Method of extraction			In all	
	Mining	In situ leaching		Area of pollution (1 000 km^2)	Volume of polluted ground (1 000 km^2)
	Area (1 000 km^2)	Area (1 000 km^2)	Volume (1 000 km^2)		
Northern ore department. Uchkuduk deposit	347.2	1 448	569	2 112.2	2 017
Eastern department. Sugraly deposit	1 060	956.6	13.4	1 121	970
Ore department No. 5 Beshkak deposit Northern and Southern Bukinai	–	–	1 923	1 740	1 923
Southern ore department. Sabyrai deposit and Ketmenchi	500	253.4	377.8	1 800	631.2
Ore department No. 2 Chauli deposit	770	–			770

The basic measures taken to lower the ecological stress in NMMC's Mining Departments are:

- *Northern ore department.* Collecting initial data for development of projects for liquidating and recultivating mining facilities that are no longer in use. A specialised organisation has started decommissioning and closure of the facilities.

- *Eastern ore department.* Technical recultivation of lands where ISL was conducted. Soil contaminated by radionuclides in a 61 000 m^2 area has been collected and deposited in the Mine No. 2 waste dump. The plan is to bury this soil in the Eastern Ore Department's radioactive waste disposal site, planned for construction in 1997.

- *Ore department No. 5.* Sanitary-radiometric research was conducted. Initial data for planning liquidation and technical recultivation for 19 of 36 ISL areas are being developed. The engineering design programme for building PZRO is developed.

- *Southern ore department.* The waste rocks with uranium content below industrial grade at the M1 Mine, Sabyrsai deposit (100 000 m^3), are being disposed of in the PZRO "Quarry". The ore dump site has been completely recultivated. Contaminated soil at the UL-1 site Sabyrsai deposit has been excavated and disposed in the PZRO "Quarry". Fifty hectares of land have been returned to local authorities. PZRO was built at the UL-4 "Gundjak" of the "Ketmenchi" deposit.

- *Ore department No. 2.* Specialised organisations conducted engineering-geological research to estimate the slope stability of the below industrial grade waste dump near the village of Krasnogorskiy. The results of the investigation indicate that part of the slope where the ore was dumped will not need to be removed.

In earlier years the Leninabadskiy Mining Chemical Combinate (LMCC) based in Leninabad, Tajikistan conducted uranium mining activities in Uzbekistan. Six of the sites have facilities and associated land that require closure and/or recultivation. Information has been obtained for some of the sites (Alatanga, Kattasai Djeindek) where decontamination and recultivation is planned.

In addition, there are formerly utilised mining facilities and sites located in the adjoining territories of Tajikistan and Kyrgyzstan that are contaminated with radioactive wastes. If they were to be disturbed they could potentially have a negative influence on the territory of Uzbekistan (tailings in the Gafurov village and waste dumps and tailings in Mailuu-Suu). A programme for cleaning up the sites and preventing the danger of radioactive contamination of lands in Kyrgyzstan and adjoining territory of Uzbekistan is now being conducted. Measures are being taken to protect the environment from radioactive contamination. First it is necessary to provide the scientists with highly-sensitive equipment to map the extent of the contaminated wastes.

Environmental considerations

More than 30 years of NMMC's uranium production related activities have impacted Uzbekistan's natural environment. This includes the areas affected by conventional mining and processing of uranium ores, as well as the operation of ISL facilities. In addition to the areas directly affected by these activities, there are surface accumulations of an estimated 2 420 thousand m^3 of sub-economic uranium bearing material. The uranium content of this material is low, ranging from 0.002 to 0.005%. This is in addition to the 60 million tonnes of tailings located near the Navoi Hydrometallurgical Plant Number 1. The total area of ground water impacted by ISL is 13 million m^2.

The related contaminated material that has been recovered from the surface area of these operations is about 3 5 million m^3.

To fully evaluate the extent of any contamination and develop a programme for reclamation and restoration, NMMC is working with Uzbekistan's leading experts and specialists from the Commonwealth of Independent States, as well as international organisations.

NMMC has developed an environmental policy for its uranium production activities. The policy is to:

- provide for the ecological safety for all NMMC facilities by using the most environmentally acceptable and cleanest in situ leach method;

- close those mining and processing companies that are economically and environmentally less effective;

- isolate and properly dispose of all accumulated radioactive waste;

- reclaim land disturbed by the company's uranium activities.

To achieve these objectives, NMMC has been developing and carrying out a step-by-step programme for evaluating and, where necessary, reclaiming the environment which may have been effected, during more than thirty years of uranium production operations.

At NMMC's hydrometallurgical plant a system of wells has been installed to monitor and control potential ground water contamination from the tailings impoundment. Recovered waters are returned to the plant for use in processing. An investigation is underway to obtain the data necessary for the selection and development of a system for covering the tailings impoundment. The first step is radioactive decontamination and reclamation of any contaminated lands surrounding the impoundment, including the pipeline route from the plant to the impoundment. Plans are also being made for covering the tailings site in the 2000 to 2005 period.

Ecological issues for ISL sites

Studies have been made of the distribution of radioactive contamination at the surface of ISL sites. They show that any contamination is localised on or just below the surface. Even for sites where significant spillage has taken place, the contamination levels return to background at a depth of 30-40 cm, because of the high sorption and neutralisation capacity of the upper soil layers.

The most significant environmental impact of ISL is on the aquifer hosting the orebody. The level of contamination of the aquifer is mostly dependant on the ISL technique used and the types of reagent and oxidising agent utilised.

The use of sulphuric acid ISL leaching systems causes an increase in the total dissolved solids in the ore body aquifer of 5 to 15 times the pre-mining values. This is accompanied by a decrease of pH from 7 to 1.5. The significant increase of mineral content in the aquifer water is primarily due to sulphate-ion accumulation in the leaching solution. The distance the leaching solution with elevated dissolved solids migrates along the flow gradient beyond the ore body limits does not exceed 150 to 200 m. When the pH increases to more than 6 this distance decreases to less than 60 m.

For bicarbonate and reagent free ISL mining, the pH, total mineral content, chemical composition and concentrations of major components in the water show practically no difference from the pre-mine ore aquifer values. This is also observed within the mining area. This applies both during and after leaching.

Data obtained from monitoring wells drilled to productive aquifers adjacent to ISL fields verify that within all of NMMC's mining areas, irrespective of which technique was used, natural geochemical water conditions beyond 200 to 300 m from the ore field limits remain unchanged. Monitoring of the aquifers above and below the productive ones also show no significant change in their natural status.

The natural pre-leach conditions of aquifers of the productive horizons of NMMC's ISL operations are regarded as not suitable for common farming and household purposes as the salt content is too high. Given this factor, the main method of eliminating aquifer contamination after ISL mining is based on natural attenuation of the residual solutions in the process of migration with the ground water stream.

Radiation protection

Radiation monitoring and safety programmes at NMMC are conducted by services for monitoring working conditions and environmental control, as well as by NMMC's industrial sanitary laboratories, regional and national sanitary inspection organisations, the National Environment Committee and the National Mining Technical Committee. In addition, a number of specialised scientific research institutions are involved in radiological studies on existing operations, as well as in pre-reclamation, post-reclamation, and methodological and operative investigations.

Results of radiation monitoring of NMMC's uranium mining and processing activities indicate that the average annual effective equivalent radiation dose to the critical population group living in these regions does not exceed 1 mSv/year for the sum of all radiation hazard factors. This is within the new basic limits determined for population by the International Commission of Radiation Protection. Many years of expertise in ISL applications by NMMC have verified the high efficiency and ecological safety of the technology when properly applied in an appropriate environment. This is expected to widen the range of the use of ISL mining.

EXAMPLE OF AN EVALUATION OF THE IMPACTS AT A URANIUM MINING AND MILLING FACILITY[1]

Summary of impacts in uranium mining and milling

Some of the various impacts that may result from uranium mining and milling are limited to certain types of operations. Table A1 lists the impacts and types of operations in which they are likely to occur.

Table A1. **Impacts and types of operations in uranium mining and milling**

Hazard	Pathway	Group at risk	Agent	Operation	ID
Radiological	Air	Workers	External Radiation	O U M	R1
			Radon Progeny	O U A B M	R2
			Skin Dose	O U M	R3
			Ingestion	O U M	R4
			Long-lived dust	O U M	R5
		Others	Radionuclide Release	O U A B M	R6
	Water	Others	Radionuclide Release	O U A B M	R7
Toxicological	Air	Worker	CO and CO_2	O U A B M	T1
			NO, Nox	O U	T2
Toxicological	Air	Worker	Diesel particulates	O U	T3
			Blasting fumes	O U	T4
			Noise	O U M	T5
			Temperature/humidity	O U A B M	T6
	Water	Others	Metals leaching	O U A B M	T7
Other	N/A	Worker	Blasting	O U	O1
			Travel/Transportation	O U A B M	O2
			Ground failure (major and minor)	O U	O3
			Fire	O U M	O4
			Vibration	U	O5
			Damp conditions	U	O6
			Heavy equipment	O U A B M	O7
			Reagents	A B M	O8
		Others	Ground opening	O U	O9
			Dam failure	M	O10
			Land alienation – temporary	O U A B M	O11
			Land alienation – permanent	O M	O12
Other	N/A	Others	Subsidence	U	O13

---

1. Taken from Appendix A of the 1996 IAEA report on *Health and Environmental Aspects of Nuclear Fuel Cycle Facilities, IAEA-TECDOC-918.*

The letters appearing in the column labelled "Operation" indicate where the impacts may occur, and have the following meanings:

O = open pit mining
U = underground mining
A = in situ leach using acid
B = in situ leach using alkaline
M = milling (including tailings management)

The various agents in Table A1 are assigned a code which is given in the column labelled "ID" for reference later in this Annex.

Evaluation of impacts

Uranium mining and milling can impact workers, the local population and the general environment. Such impacts can be of three types: radiological, toxicological and other. The significance of these impacts for a particular mine or mill is controlled, primarily, by four factors: the type of operation, the grade of ore being mined or milled, the age of the facility (since older facilities do not, in general, have the same radiation and environmental protection controls available), and the density of the local population. Combining the factors leads to the possibility of 40 different categories of facilities that may exist. In practice, however, not all factors are significant in all cases, and the number of cases is less.

It is impossible to develop a single model that is representative of all categories, as there are major differences that cannot be averaged when determining the impacts. However, it is theoretically possible to develop a model for each category, and then to determine the total impact of uranium mining and milling operations world-wide by summing the impacts category, weighted by the amount of uranium that comes from the category.

To make this task manageable, Table A2 has been developed. This table identifies the impacts that may be significant under a particular set of circumstances. In order to determine the total impact, only those impacts that are significant need be evaluated. While other impacts may be present, their contribution to the total impact is likely to be small.

The assumption is made that appropriate management practices are used to limit certain impacts and, therefore, such impacts are not included. Should this not be the case, it will be necessary to consider all impacts that may result.

It should be noted that the impacts on the local population and on the environment are likely to be different during operations and after closure of the facility. The entries under milling include the possible impacts from the tailings management facility during operations and after closure.

The entries for in situ leaching include the possible impacts from both the mining and processing. For conventional mining and milling operations, it will generally be necessary to consider the impacts from mining and milling independently.

The letters and numbers appearing in the columns labelled "Significant Impacts" indicate the identification of the impact that may occur, as listed in the column marked "ID" in Table A1. Where there is no entry, no significant impacts are anticipated.

It is not intended that the differentiation between high and low grade, between a new facility and an old one, and between dense population and sparse population in the vicinity of the facility be absolute. As a general guide in interpreting the table, the following definitions may be used:

High grade: uranium content of the ore greater than 0.5%.
Low grade: uranium content of the ore less than 0.5%.
New facility: facility constructed to currently accepted environmental and safety standards.
Old facility: other facilities.
Dense population: an average of greater than 3 people per km^2 living within 25 kilometres.
Sparse population: an average of less than 3 people per km^2 living within 25 kilometres.

Table A2. **Potential impacts to workers, local population and environment**

Operation Type	Grade of ore	Age of facility	Local population	Significant Impacts				
				Workers	Local population		Environment	
				During	*During*	*After*	*During*	*After*
Open pit	High	New	Dense	R1,2,5 T2,4,5,6 O1,2,3,7	R6 O11	R7 T7 O9	R6,7 T7	T7
			Sparse			O9		
		Old	Dense	R1,2,5 T2,4,5,6 O1,2,3,7	R6 T7 O11	R7 T7 O9	R6,7 T7	R7 T7
			Sparse		R6 T7	O9		
	Low	New	Dense	T2,4,5,6 O1,2,3,7	R6 T7 O11	T7 O9	T7	T7
			Sparse			O9		
		Old	Dense	T2,4,5,6 O1,2,3,7	R6 O11	T7 O9	R7 T7	R7 T7
			Sparse			O9		
Under-ground	High	New	Dense	R1,2,3,5 T2,3,4,5 O1,2,3,4,5,6,7	R6 O11	O13	R6	
			Sparse					
Under-ground	High	Old	Dense	R1,2,3,5 T2,3,4,5 O1,2,3,4,5,6,7	R6 O11	O13	R6	R7
			Sparse					
	Low	New	Dense	R1,2,5 T2,3,4,5 O1,2,3,4,5,6,7	R6 O11	O13		
			Sparse					
		Old	Dense	R1,2,5 T2,3,4,5 O1,2,3,4,5,6,7	R6 O11	O13	R6	
			Sparse					
In situ leaching – acid	High or low	New or old	Dense or sparse	T6 O2,7,8	O11	R7 T7		R7 T7

157

Operation Type	Grade of ore	Age of facility	Local population	Significant Impacts				
				Workers	Local population		Environment	
				During	During	After	During	After
In situ leaching – alkaline	High or low	New or old	Dense or sparse	T6 O2,7	O11			
Milling	High	New	Dense	R1,2,3,4,5 T5,6 O2,4,7,8	R6,7 O11	O10,12	R6,7	O10
			Sparse			O12		
		Old	Dense	R1,2,3,4,5 T5,6 O2,4,7,8	R6,7 O11	R7 T7 O11	R6,7	R7 T7 O10
			Sparse			R7 T7 O12		
	Low	New	Dense	R1,2,4,5 T5,6 O2,4,7,8	O11	O12		O10
			Sparse			O12		
Milling	Low	Old	Dense	R1,2,4,5 T5,6 O2,4,7,8	R6,7 O11	O12	R7	R7 T7 O10
		Old	Sparse			O12		

An illustrative example

Mine: Key Lake, Canada
Type: Open pit and mill
Grade of ore: High
Age of facility: New (1983)
Local population: Sparse
Production: 5 314 tonnes of uranium in 1993 (16.3% of world production)

Possible significant impacts of the mine

R1:   External radiation exposure of workers
R5:   Long-lived radioactive dust exposure of workers
R6:   Impact on the environment of radionuclide releases to the air during operations
R7:   Impact on the environment of radionuclide releases to the aquatic environment during operations
T2:   NO, NOx exposure of workers
T4:   Blasting fume exposure of workers
T5:   Noise exposure of workers
T6:   Temperature and humidity exposure of workers
T7:   Impact on the environment of metals leaching into the aquatic environment during operations and after the completion of operations
O1:   Blasting accidents to workers
O2:   Travel and transportation accidents to workers

O3: Ground failure accidents to workers
O7: Heavy equipment accidents to workers
O9: Accident risk resulting from the existence of the open pit after completion of operations

Possible significant impacts of the mill

R1:   External radiation exposure of workers
R2:   Radon progeny exposure of workers
R3:   Radiation skin dose of workers
R4:   Ingestion of radioactive material
R5:   Long-lived dust exposure of workers
R6:   Impact on the environment of radionuclide releases to the air during operations
R7:   Impact on the environment of radionuclide releases to the aquatic environment during operations
T5:   Noise exposure of worker
T6:   Temperature and humidity impact on workers
O2:   Travel and transportation accidents to workers
O4:   Fire occurrence
O7:   Heavy equipment accidents to workers
O8:   Accidents to workers involving reagents
O10: Probability and environmental consequences of a dam failure after completion of operations
O12: Amount of land that will be alienated from future use

IAEA ACTIVITIES AND PUBLICATIONS RELEVANT TO ENVIRONMENTAL ISSUES IN URANIUM MINING AND MILLING

Environmental and safety issues related to uranium mining and milling are the responsibility of two divisions of the IAEA. The relevant programmes are managed by the Division of Radiation and Waste Safety (NSRW) and the Division of Nuclear Fuel Cycle and Waste Technology (NEFW).

The relevant activities are conducted through the regular IAEA programme, as well as the programme of technical co-operation with developing countries. The regular programme involves promoting information exchange through meetings and publications. It also includes the recognition, definition and promotion of international standards and guidelines. These activities are accomplished through advisory groups, technical committees and consultants meetings. Technical co-operation involves technology transfer activities which is accomplished through projects, fellowship training and scientific visits, as well through expert missions to advise.

The NSRW Division develops documents in the form of safety standards. The documentation for standards from this Division is reviewed by Standard Advisory Committees with participants from Member countries. Currently, the Division is preparing a Safety Guide Series Publications with the title *Safe Management of Radioactive Waste from Mining and Milling of Uranium and Thorium Ores*.

The NEFW Division is involved in technical co-operation projects and has recently established the UPSAT Environmental Programme and Safety Evaluation Mission. Planning is underway to conduct an evaluation of one mine/mill operation under this programme. This Division is also conducting a regional project in Central and Eastern Europe with the main focus of identifying and characterising radioactively contaminated sites in the region.

The NEFW Division conducted a regional training course in uranium mining, including environmental aspects and is carrying out a Technical Co-operation Model Project entitled "Modern Technologies for In Situ Leach Uranium Mining" in Kazakhstan. This project includes a baseline study, environmental impact assessment, good practice in design, operation and restoration, as well as other topics related to in situ leach mining. Another project "Treatment of liquid effluents from uranium mines and mills during and after operation (post decommissioning rehabilitation)" is conducted as a co-ordinated research project with participation from 11 countries. The Division has conducted other activities related to sound environmental practices for in situ leach uranium mining.

Technical co-operation projects and publications

The IAEA has a long history of publications and technical co-operation (TC) projects in the field of uranium mining and milling waste. In the late 1980s, TC projects in this field were successfully implemented among others in Argentina and Portugal. An IAEA TC mission in 1992 to the Rössing Uranium Mine in Namibia disclosed that despite limited resources, dedicated efforts of the management yielded highly satisfactory results in achieving acceptably low levels of contamination and prevention of spread of radioactive contaminants. IAEA sponsored assessments of the radiological conditions at mines and mills in several other countries have shown that uranium production can be carried out in a safe manner without detriment to the environment.

A list of IAEA publications related to environmental issues in uranium mining and milling is included in the last section of this annex. A report published in 1992 by the IAEA entitled "Current Practices for the Management of Uranium Mill Tailings" (Technical Reports Series No. 335) discusses the current practices used in the design, siting, construction and operation of impoundment facilities for uranium mill tailings. The report presents an integrated overview of the technological safety and radiation protection aspects in order to ensure that the potential radiological and non-radiological risks associated with the management of uranium mill tailings are minimised. This report:

- Identifies the nature and source of radioactive and non-radioactive pollutants in uranium mill tailings.

- Identifies the important mechanisms by which pollutants can be released from the tailings impoundment and the variables that control these mechanisms.

- Reviews radiation protection aspects of these mechanisms.

- Describes the pathways by which the pollutants may reach humans.

- Describes some of the site selection and design options that may be implemented to facilitate disposal and/or control the extent of releases from the impoundment.

Aspects related to decommissioning of facilities for and close-out of residues from mining and milling of uranium ores are emphasised in another technical report published in 1994 by the IAEA entitled "Decommissioning of Facilities for Mining and Milling of Radioactive Ores and Close-out of Residues" (Technical Reports Series No. 362). This report presents an overview of factors involved in planning and implementation of decommissioning/close-out of uranium mine/mill facilities. The information contained applies to mines, mills, tailings piles, mining debris piles and leach residues that are present as operational, mothballed or abandoned projects, as well as to future mining and milling projects. The report identifies the major factors that need to be considered in the decommissioning/close-out activities, including regulatory considerations; decommissioning of mine/mill buildings, structures and facilities; decommissioning/close-out of open pit and underground mines; decommissioning/close-out of tailings impoundments; decommissioning/close-out of mining debris piles, unprocessed ore and other contaminated material such as heap leach piles, in situ leach facilities and contaminated soils; restoration of the site, vicinity properties and groundwater; radiation protection and health and safety considerations; and an assessment of costs and post-decommissioning or post close-out maintenance and monitoring aspects.

Another document entitled "A Review of Current Practices for the Close-out of Uranium Mines and Mills" is being published by the IAEA. This is a world-wide overview of national strategies/practices for decommissioning/close-out of uranium mining and milling facilities and sites. The aim is to identify typical issues and national approaches to their solutions.

The recent political changes in Central and Eastern Europe (CEE) and the former Soviet Union have revealed the state of the environment in these countries. An extensive industrial build up and heavy depletion of natural resources in these countries to accomplish the quota based productivity goals have resulted in neglect of the preservation and protection of the environment.

The radioactive contaminant materials resulting from diverse activities related to the nuclear fuel cycle, defence and industries, in addition to medical and research applications have taken a heavy toll on the environment causing severe pollution, a legacy from the past era. The waste resulting from some fifty years of activities has piled up – often at unregistered sites – leaving the public with the potential danger of long-term radiation exposure.

The political changes not only brought forward a fragmentary disclosure of the nuclear contaminated sites, but also resulted in a condition in which these countries became receptive to co-operation from a category of countries the region had previously been isolated from.

It was in these circumstances of change that the IAEA decided to launch the Technical Co-operation Regional Project RER/9/022 on Environmental Restoration in Central and Eastern Europe. The first phase of the project was initiated in the latter part of 1992 and concluded in 1994.

The main focus of the project was the identification and characterisation of radioactively contaminated sites in the region of CEE. Before any action regarding environmental restoration could be taken, the countries involved and the IAEA needed to obtain an overview of the environmental status in each of the countries. The Agency requested selected experts from the targeted Member states to present a comprehensive report identifying contaminated sites in their home countries. The experts have, as thoroughly as possible, surveyed and categorised each site according to location, volume and radioactive concentration. Also included were data on radioactive concentrations in the sites, source(s) of contamination, description of radiological hazard (e.g. proximity to population areas), potential for spreading of contamination, etc. Wherever possible, organisations responsible for the monitoring and clean-up of each of the contaminated sites have also been identified. The results of the 1993-94 phase were published in a technical report by the IAEA in 1996, entitled "Planning for Environmental Restoration of Radioactively Contaminated Sites in Central and Eastern Europe" (IAEA-TECDOC 865, 3 volumes).

During implementation of the IAEA project, it became apparent that most countries in the region share the problem of contamination from uranium mining and milling (accident-generated contamination involves relatively fewer countries; and other TC projects deal with these matters). As a follow-up, a second phase was therefore established for 1995-1996 with the aim of focusing on radioactive contamination from uranium mining and milling and the development of plans for environmental restoration.

While the 1993-1994 phase aimed at attracting the attention of Member states to a long neglected problem, it was felt that the time has come for Member states to initiate concrete planning activities that would lead to corrective actions in highly contaminated areas. The emphasis shifted from scientific discussion to the identification of responsibilities, planning activities, and assessment of existing and needed resources for the eventual implementation of restoration plans.

The 1995-96 phase included a preliminary planning meeting and three workshops. The planning meeting, involving designated experts from the region, aimed at determining the expected inputs from the Member states of the region. It should be noted that two years was not enough to allow detailed elaboration of environmental restoration plans. However, preliminary plans were expected by the end of the project. Based on the results of the planning meeting, subsequent workshops were conducted in parallel with planning activities in Member states. During workshops, designated experts presented progress reports. Countries with hands-on experience in environmental restoration provided practical information and know-how. The Project schedule included a planning meeting in Vienna (March 1995), Workshops in Bulgaria (October 1995), Ukraine (April 1996) and the latest one in Romania (November 1996).

The second objective of the Vienna meeting was to establish the implementation mechanisms of future meetings within the Project. Following the discussions, it was decided to use three mechanisms in parallel at any given workshop:

- Peer review of selected environmental restoration projects/studies. This mechanism was ideally implemented when the site subject to peer review was visited by the group of designated experts. To improve efficiency of the peer review process, the host country sent relevant information to the experts well before the workshop when the site in question was to be visited. The scope and extent of the peer review process depend on the information provided by the host country. For example, at the workshop in Sofia, Bulgaria in October 1995, such a mechanism was successfully implemented for the Bukhovo (near Sofia) environmental restoration project.

- Progress reports. Member states participating in the IAEA project are developing environmental restoration plans and studies which were presented in progress reports.

- Discussion on specific topics. Such topics are assigned to specified experts from and/or outside CEE for workshop presentation. Some of the topics dealt with in the project have been as follows: radiological characterisation and dose assessment; clean-up standards and other regulations; groundwater contamination and remediation; and costs.

The following countries were primarily addressed by the project: Bulgaria, the Czech Republic, Estonia, Kazakhstan, Kyrgyzstan, Poland, Romania, Slovenia, Ukraine and Uzbekistan. Designated experts from these countries took part in the above-mentioned workshops. Invited speakers from the following countries: Canada, France, Germany, Russian Federation, Spain and the USA provided information on planning for and implementing environmental restoration based on their national experience. The main results of the 1995-96 phase of this TC project are being prepared for publication by the IAEA. A TECDOC will collect papers submitted by national experts during and after the Romanian workshop in November 1996. Papers describe national progress achieved over the time frame of the project, problems encountered and short- to medium-term prospects. Following the completion of the above-described Regional project, several national Technical Co-operation projects started in 1997-98. In Eastern Europe, they include environmental restoration of Mining and Milling facilities in Bulgaria, Czech Republic and Slovenia. Outside Europe, one project is being implemented in China.

Future activities include a co-ordinated research project on "Technologies and Methods for Long-Term Stabilisation and Isolation of Uranium Mill Tailings". In addition, a Technical Report on "Monitoring and Surveillance of Uranium Mining and Milling Tailings", will be prepared by the Division of Radiation and Waste Safety in the context of the RADWASS Programme.

Furthermore, technical documents (TECDOCS) on the following topics are in preparation or planned:

- Technologies for the Remediation of Contaminated Sites.

- Technical Options for the Remediation of Contaminated Groundwaters.

- Post-Restoration Monitoring of Decommissioned Sites to Ensure Compliance with Clean-up Criteria.

- Site Characterisation Techniques Used in Environmental Restoration.

- A World-Wide Directory of Radioactively Contaminated Sites.

- Environmental Restoration Practices as determined by land-use, cost, public perception and other factors impacting on the decision-making process (Decision-Making System).

- "The Impact of New Environmental and Safety Regulations on Uranium Exploration, Mining, Milling and Waste Management", proceedings of an IAEA Technical Committee Meeting (TCM), 14-17 September 1998, Vienna, Austria.

IAEA Publications relating to the subject area of uranium/thorium mining and milling

Safety fundamentals

IAEA (1995), RADWASS Safety Fundamentals – *The Principles of Radioactive Waste Management*, Safety Series No. 111-F, Vienna, Austria.

Safety standards

IAEA (1995), RADWASS Safety Standard on *Establishing a National Legal System for Radioactive Waste Management*, Safety Series No. 111–S–1, Vienna, Austria.

IAEA (1996), International Basic Safety Standards for *Protection against Ionising Radiation and for the Safety of Radiation Sources*, Safety Series No. 115, Vienna, Austria.

IAEA (1983), *Radiation Protection of Workers in the Mining and Milling of Radioactive Ores*, Safety Series No. 26, Vienna, Austria.

IAEA (1987), *Safe Management of Wastes from the Mining and Milling of Uranium and Thorium Ores* – Code of Practice and Guide to the Code, Safety Series No. 85, Vienna, Austria.

IAEA (1996), *Regulations for the Safe Transport of Radioactive Material*, Safety Standards Series No. ST-1, Vienna, Austria.

Safety guides

IAEA (1989), *Radiation Monitoring in the Mining and Milling of Radioactive Ores*, Safety Series No. 95, Vienna, Austria.

IAEA (1994), *Classification of Radioactive Waste*, Safety Series No. 111-G-1.1, Vienna, Austria.

IAEA, *Safe Management of Radioactive Waste from the Mining and Milling of U/Th Ores*, Safety Series No. NS–277, Vienna, Austria (in preparation).

IAEA (1987), *Application of the Dose Limitation System to the Mining and Milling of Radioactive Ores*, Safety Series No. 82, Vienna, Austria.

Safety practices

IAEA, *Monitoring and Surveillance of Mining and Milling Tailings*, Procedures for Close-out of Mines, Waste Rock and Mill Tailings, Safety Series No. NS–XXX, Vienna, Austria (in preparation).

Procedures and data

IAEA (1989), *The Application of the Principles for Limiting Releases of Radioactive Effluents in the Case of the Mining and Milling of Radioactive Ores*, Safety Series No. 90, Vienna, Austria.

Technical reports series

IAEA (1994), *Decommissioning of Facilities for Mining and Milling of Radioactive Ores and Close-out of Residues*, Technical Reports Series No. 362, Vienna, Austria.

IAEA (1992), *Current Practices for the Management and Confinement of Uranium Mill Tailings*, Technical Reports Series No. 335, Vienna, Austria.

IAEA (1992), *Measurement and Calculation of Radon Releases from Uranium Mill Tailings*, Technical Reports Series No. 333, Vienna, Austria.

IAEA (1994), *Assessment and Comparison of Waste Management System Costs for Nuclear and Other Energy Sources*, Technical Reports Series No. 366, Vienna, Austria.

Other reports

IAEA (1992), *Radioactive Waste Management*, An IAEA Source Book, Vienna, Austria.

IAEA (1986), *Environmental Migration of Radium and Other Contaminants Present in Liquid and Solid Wastes from the Mining and Milling of Uranium*, IAEA-TECDOC 370, Vienna, Austria.

IAEA (1994), *Safety Indicators in Different Time Frames for the Safety Assessment of Underground Radioactive Waste Repositories*, IAEA-TECDOC 767, Vienna, Austria.

Gnugnoli, G., Laraia, M., Stegnar, P., *Uranium Mining and Milling: Assessing Issues of Environmental Restoration*, IAEA Bulletin, Vol. 38, No. 2, 1996, pp. 22-26, Vienna, Austria.

Recommendations and guidelines

IAEA (1975), *Radon In Uranium Mining*, Panel Proceedings Series, Vienna, Austria.

IAEA (1981), *Current Practices and Options for Confinement of Uranium Mill Tailings*, Technical Reports Series No. 209, Vienna, Austria.

IAEA (1990), *The Environmental Behaviour of Radium*, Technical Reports Series No. 310, Vienna, Austria.

IAEA (1991), *Airborne Gamma Ray Spectrometer Surveying*, Technical Reports Series No. 323, Vienna, Austria.

IAEA (1992), *Analytical Techniques in Uranium Exploration and Ore Processing*, Technical Reports Series No. 341, Vienna, Austria.

IAEA (1982), *Management of Wastes from Uranium Mining and Milling*, Proceedings Series, Vienna, Austria.

IAEA (1989), *In Situ Leaching of Uranium: Technical, Environmental and Economic Aspects*, TECDOC–492, Vienna, Austria.

IAEA (1993), *Uranium In Situ Leaching*, TECDOC–720, Vienna, Austria.

IAEA (1995), *Guidelines for Comparative Assessment of the Environmental Impacts of Wastes from Electricity Generation Systems*, TECDOC 787, Vienna, Austria (includes impact from uranium mining).

IAEA (1995), *Planning and Management of Uranium Mine and Mill Closures*, TECDOC–824, Vienna, Austria.

IAEA (1996), *Guidebook on the Development of Regulations for Uranium Deposit Development and Production*, TECDOC–862, Vienna, Austria.

ANNEX 3

MEMBERS OF THE JOINT NEA-IAEA URANIUM GROUP

Argentina	Mr. A. CASTILLO	Comisión Nacional de Energía Atómica Unidad de Proyectos Especiales de Suministros Nucleares, Buenos Aires
Australia	Mr. I. LAMBERT Mr. A. McKAY	Department of Primary Industries and Energy, Bureau of Resource Sciences Canberra
	Mr. R. JEFFREE	Australian High Commission London
Belgium	Ms. F. RENNEBOOG	Synatom, Brussels
Brazil	Ms. E.C.S. AMARAL	Instituto de Radioproteçâo e Dosimetria (CNEN/IRD), Rio de Janeiro
	Mr. S. SAAD	Comissâo Nacional de Energia Nuclear (COMAP/CNEN), Rio de Janeiro
	Mr. H.A. SCALVI	Indústrias Nucleares do Brasil S/A – INB Poços de Caldas
Canada	Mr. R.T. WHILLANS	Uranium and Radioactive Waste Division Natural Resources Canada Ottawa
China	Mr. R. ZHANG	Bureau of Mining and Metallurgy China National Nuclear Corporation (CNNC), Beijing
Czech Republic	Mr. J. ŠURÁN **Vice-Chairman Uranium Group** Mr. J. BADAR Mr. J. MAKOVICKA	DIAMO s.p. Stráz pod Ralskem

Czech Republic *(contd.)*	Mr. R. MAYER	Ministry of Industry and Trade Prague
Egypt	Mr. A.B. SALMAN	Nuclear Materials Authority (NMA) El-Maadi, Cairo
Finland	Dr. K. PUUSTINEN	Department of Economic Geology Geological Survey of Finland Espoo
France	Mr. J-L. BALLERY **Vice-Chairman Uranium Group** Mrs. F. THAIS	Commissariat à l'Énergie Atomique Centre d'Études de Saclay Saclay
Germany	Dr. F. BARTHEL **Chairman Uranium Group**	Bundesanstalt für Geowissenschaften und Rohstoffe, Hannover
Greece	Mr. D.A.M. GALANOS	Institute of Geology and Mineral Exploration, Athens
Hungary	Mr. G. ÉRDI-KRAUSZ **Co-Chairman Working Group**	Mecsekuran Ltd. Pécs
India	Dr. C.K. GUPTA	Bhabha Atomic Research Centre Mumbai, Bombay
	Mr. K.K. DWIVEDY	Atomic Minerals Division Department of Atomic Energy Hyderabad
Iran	Mr. A.G. GARANKANI Mr. S.M.R. AYATOLLAHI Mr. M.H. MASHAYEKHI	Atomic Energy Organisation of Iran Tehran
Japan	Mr. H. MIYADA	Geotechnics Development Section Tono Geoscience Centre Power Reactor and Nuclear Fuel Development Corp. (PNC), Toki-shi Tokyo
	Mr. K. NORITAKE	PNC Paris Office, Paris

Jordan	Mr. S. AL-BASHIR	Jordan Phosphate Mines Company Amman
Kazakhstan	Mr. G.V. FYODOROV	Atomic Energy Agency of Kazakhstan Almaty
Lithuania	Mr. K. ZILYS	Acting Resident Representative of Lithuania, Vienna, Austria
Mongolia	Mr. T. BATBOLD	Uranium Co., Ltd. Ulaanbaatar
Morocco	Mr. D. MSATEF	Centre d'Études et de Recherches des Phosphates Minéraux Casablanca
Namibia	Mr. H. ROESENER	Geological Survey Ministry of Mines and Energy Windhoek
Netherlands	Mr. J.N. HOUDIJK	Ministry of Economic Affairs The Hague
Pakistan	Mr. M.Y. MOGHAL	Atomic Energy Minerals Centre Lahore
Philippines	Ms. P.P. GARCIA	Philippine Embassy Pretoria
Portugal	Mr. R. DA COSTA	Instituto Geologico e Mineiro Lisbon
Russian Federation	Mr. A.V. BOITSOV	All-Russian Research Institute of Chemical Technology, Moscow
	Mr. S.S. NAUMOV	Geologorazvedka Moscow

Russian Federation (*contd.*)	Mr. A.V. TARKHANOV	Ministry of the Russian Federation on Atomic Energy, Moscow
South Africa	Mr. B.B. HAMBLETON-JONES **Vice-Chairman Uranium Group** Mr. L.C. AINSLIE Mr. R.G. HEARD	Atomic Energy Corporation of South Africa Ltd. Pretoria, South Africa
Spain	Mr. J. ARNÁIZ DE GUEZALA	Empresa Nacional del Uranio S.A. (ENUSA), Madrid
Sweden	Dr. I. LINDHOLM **Chairman Working Group**	Swedish Nuclear Fuel & Waste Management Co. Stockholm
Switzerland	Mr. R.W. STRATTON	Nordostschweizerische (NOK) Kraftwerke AG Baden
Turkey	Mr. Z. ERDEMIR	Turkish Electricity Generation Ankara
Ukraine	Mr. A.Ch. BAKHARZHIEV	The State Geological Company "Kirovgeology", Kiev
	Mr. A.P. CHERNOV	The Ukrainian State Committee on Nuclear Power Utilisation, (Goscomatom) Kiev
	Mr. B.V. SUKHOVAROV-JORNOVYI	Scientific, Technological and Energy Centre Kiev
United Kingdom	Mr. N. JONES	Rio Tinto Mineral Services Ltd. London
United States	Mr. J. GEIDL **Vice-Chairman Uranium Group**	Energy Information Administration US Department of Energy Washington
	Mr. W. FINCH	US Geological Survey Denver

Uzbekistan	Mr. N.S. BOBONOROV	State Committee on Geology and Mineral Resources of the Republic of Uzbekistan Tashkent
	Mr. S.B. INOZEMTSEV	Navoi Mining and Metallurgy Combinat Navoi
European Commission	Mr. J-P. LEHMANN	Directorate General XVII (Energy) Nuclear Energy Brussels, Belgium
IAEA	Dr. D.H. UNDERHILL **Scientific Secretary**	Division of Nuclear Fuel Cycle and Waste Technology Vienna, Austria
OECD/NEA	Dr. I. VERA **Scientific Secretary**	Nuclear Development Division Paris, France

CONTRIBUTORS TO DRAFTING AND REVIEW OF THE REPORT

Mr. S. NEEDHAM	Department of the Environment, Australia
Mr. M.I. RIPLEY	Counsellor (Nuclear), Australian Embassy, Vienna
Mr. T. CHUNG	Energy Information Administration US Department of Energy

ANNEX 4

LIST OF REPORTING ORGANISATIONS

Argentina

Comisión Nacional de Energía Atómica, Unidad de Proyectos Especiales de Suministros Nucleares, Avenida del Libertador 8250, 1429 Buenos Aires

Australia

Bureau of Resource Sciences, P.O. Box E 11, Kingston ACT 2604

Office of the Supervising Scientist, Level 2, Tourism House, 40 Blackal Street, Barton ACT 2600

Brazil

Instituto de Radioprotecaô e Dosimetrica, Av. Salvador Allende, S/N – Via 9, CP 37750, 22780 – 160 Rio de Janeiro

Industrias Nucleares Do Brazil S/A - INB, P.O. Box 961, 37701-970 Poços de Caldas – MG

Canada

Uranium and Radioactive Waste Division, Natural Resources Canada, 580 Booth Street, Ottawa, Ontario K1A OE4

China

Bureau of Mining and Metallurgy, China National Nuclear Corporation (CNNC), P.O. Box 2102-9, Beijing 100822

Czech Republic

DIAMO s.p., 47127 Stráz pod Ralskem

Finland

Geological Survey of Finland, P.O. Box 96, FIN-02151 Espoo

France

Commissariat à l'Énergie Atomique, Centre d'études nucléaires de Saclay, F-91191-Gif-sur-Yvette Cédex

Gabon

Ministère des Mines, de l'Énergie et du Pétrole, B.P. 874, Libreville

Germany

Bundesanstalt für Geowissenschaften und Rohstoffe, Stilleweg 2, D-30655 Hannover

Hungary

Mecsekurán LLC, P.O. Box 65, Kövágószölös 0165, H-7614 Pécs

India

Atomic Minerals Division, Department of Atomic Energy, 1-10-153-156, Begumpet, Hyderabad 500 016

Japan

JPNC Paris Office, 4-8 rue St Anne, F-75001 Paris

Jordan	Natural Resources Authority, P.O. Box 7, Amman
	Jordan Phosphate Mines Co., P.O. Box 30, Amman
Kazakhstan	Atomic Energy Agency, 13 Republic Square, Almaty, 480013
Namibia	Geological Survey of Namibia, P.O. Box 2168, Windhoek
Niger	Ministère des Mines et de l'Énergie, Direction des Mines, B.P. 11700, Niamey
Portugal	Ministério da Indústria e Energia, Instituto Geológico e Mineiro, Rua Almirante Barroso, 38, P-1000 Lisbon
Russian Federation	All-Russian Institute of Chemical Technology, Ministry of Atomic Energy, 33 Kashirskoye Shosse, 115230 Moscow
South Africa	Atomic Energy Corporation, P.O. Box 582, Pretoria 0001
Spain	ENUSA, Santiago Rusiñol 12, E-28040 Madrid
Sweden	Swedish Nuclear Fuel and Waste Management Co. (SKB), P.O. Box 5864, S-102 40 Stockholm
	Studsvik Eco & Safety AB, S-611 45 Nyköping
Ukraine	State Committee on Nuclear Power Utilization (Goscoatom), Arsenalnaya str. 9/11, Kiev 252011
United States	Energy Information Administration, Coal, Nuclear, Electric and Alternate Fuels (EI- 50), U.S. Department of Energy, Washington, D.C. 20585
Uzbekistan	The State Geological Company "Kyzyltepageologia", 7a, Navoi Street, 700000 Tashkent

OECD PUBLICATIONS, 2, rue André-Pascal, 75775 PARIS CEDEX 16
PRINTED IN FRANCE
(66 1999 09 1 P) ISBN 92-64-17064-2 – No. 50875 1999